A Zeptospace Odyssey

A Zeptospace Odyssey

A Journey into the Physics of the LHC

Gian Francesco Giudice

CERN, Department of Theoretical Physics
Geneva, Switzerland

OXFORD
UNIVERSITY PRESS

OXFORD

UNIVERSITY PRESS

Great Clarendon Street, Oxford OX2 6DP

Oxford University Press is a department of the University of Oxford.
It furthers the University's objective of excellence in research, scholarship,
and education by publishing worldwide in

Oxford New York

Auckland Cape Town Dar es Salaam Hong Kong Karachi
Kuala Lumpur Madrid Melbourne Mexico City Nairobi
New Delhi Shanghai Taipei Toronto

With offices in

Argentina Austria Brazil Chile Czech Republic France Greece
Guatemala Hungary Italy Japan Poland Portugal Singapore
South Korea Switzerland Thailand Turkey Ukraine Vietnam

Oxford is a registered trade mark of Oxford University Press
in the UK and in certain other countries

Published in the United States
by Oxford University Press Inc., New York

© Gian Francesco Giudice 2010

The moral rights of the authors have been asserted
Database right Oxford University Press (maker)

First published 2010
Reprinted 2010

British Library Cataloguing in Publication Data

Data available

Library of Congress Cataloging in Publication Data

Data available

Typeset by SPI Publisher Services, Pondicherry, India
Printed in Great Britain
on acid-free paper by
CPI Antony Rowe, Chippenham, Wiltshire

ISBN 978–0–19–958191–7 (Hbk.)

3 5 7 9 10 8 6 4 2

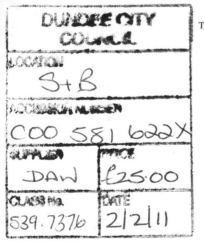

Contents

1
Prologue

We shall not cease from exploration
And the end of all our exploring
Will be to arrive where we started
And know the place for the first time.

Thomas Stearns Eliot[1]

The control room of the Large Hadron Collider (LHC) is packed full of people. Everyone's eyes are fixed on the monitor screen hanging on the wall, which now is showing only a grey background. The last absorber block has already been removed, and so the protons will not find any obstacle to their circular trajectory around the 27-kilometre long underground tunnel. It is 10.28 a.m. on 10 September 2008. We are at CERN, the European research laboratory for particle physics, stretched across the border between France and Switzerland, near Geneva.

Lyn Evans, the director of the LHC project, like a magician ready to perform his most astonishing trick, recites the magic formula in French, but without hiding his lilting Welsh intonation: "Trois, deux, un . . . . . . faisceau!" At that very moment, two white spots appear for an instant on the screen. Everyone bursts into applause. The images from the control room are broadcast live into the main auditorium, where most of the CERN physicists and staff have gathered. There too, everyone joins in spontaneous applause, full of satisfaction and emotion. The adventure, so long worked for and waited for, has really started.

The first official studies for the LHC, the most powerful particle accelerator in the world, date from the early 1980s, but the project was finally approved only in 1994. Fourteen years later, those two small white spots on the screen marked the end of the construction phase and the beginning of the experimental programme in particle physics. Those spots were in fact the two images left on a thin fluorescent film by the proton beam. One spot showed the beam at the instant it was injected

---

[1] T.S. Eliot, *The Four Quartets,* Harcourt, New York 1943.

Figure 1.1 The LHC control room on 10 September 2008.
Source: CERN.

into the LHC; the second spot was left by the beam at the instant it returned there, after circulating once around the ring, covering 27 kilometres in only 90 millionths of a second. It is true that the proton beam energy was only a small fraction of what it will be when the LHC runs at full power. Also, the density of circulating protons was extremely low. However, the sincere applause of the physicists gathered for the event is well justified, because this was the decisive test that the technology on which the LHC is based really does work.

In the control room, the last five CERN director generals are present. They are the men who led the laboratory during the various phases of LHC planning and construction: Herwig Schopper, Carlo Rubbia, Christopher Llewellyn Smith, Luciano Maiani, and finally Robert Aymar, whose mandate expired at the end of 2008 and who was succeeded by Rolf Heuer. "Only five are present because the others are already dead!" comments Lyn Evans with jovial laughter. Some of the older directors do not share the hilarity. However, visibly delighted, in their suits and ties, they approach Evans, dressed in more traditional CERN attire – jeans and sneakers – to express their joy. Congratulation messages reach CERN from all the main laboratories for particle physics around the world. The most original one is from Nigel Lockyer, director of the Canadian laboratory TRIUMF who writes, rephrasing Neil Armstrong's words on first setting foot on the moon: "One short trip for a proton, but one giant leap for mankind!"

The LHC represents indeed an extraordinary adventure for mankind. It is a demanding civil engineering adventure with, for instance, excavation of an artificial cavern of almost 80 000 cubic metres at 100 metres below ground – large enough to fill the nave of Canterbury cathedral. It is an adventure at the absolute forefront of technology, with the development of new materials and innovative instruments. It is an unprecedented adventure in information technology, with a flux of data of about a million gigabytes per second – the same as about ten phone calls placed simultaneously through the same operator by every single inhabitant of the earth. But, above all, it is a fantastic intellectual adventure, because the LHC will explore spaces where no previous experiment has ever been able to penetrate. The LHC is a journey inside the deepest structure of matter, aiming at the discovery of the fundamental laws that determine the behaviour of nature. At stake is the understanding of the first principles that govern the universe, of how and – especially – of why nature works in the way we see it operating.

The most fascinating aspect of the LHC is its journey towards the unknown. The LHC acts as a gigantic microscope able to peer at distances less than about 100 zeptometres. The zeptometre is a rarely used unit equal to a billionth of a billionth of a millimetre. The term was invented in 1991 by the Bureau International des Poids et Mesures with the motivation: "the prefix 'zepto' is derived from 'septo' suggesting the number seven (the seventh power of 1000) and the letter 'z' is substituted for the letter 's' to avoid the duplicate use of the letter 's' as a symbol."[2] This is quite an odd definition for an odd unit of measurement. Everything about the word "zepto" is so odd that I find it very appropriate to describe the unknown and strange space of extremely small distances. This infinitesimal space, no larger than a few hundred zeptometres, has been accessed so far only by elementary particles and by the wild imagination of theoretical physicists. In this book this space will be referred to as *zeptospace*. The LHC will be the first machine to explore zeptospace.

While the manned mission to the moon had a concrete goal – visible to everyone on any cloudless night – the journey that the LHC has begun is an odyssey towards stranger spaces in which no one can predict exactly what we will meet or where we will arrive. It is a search for unknown worlds which is carried out with complex cutting-edge technologies and guided by theoretical speculations whose understanding requires knowledge of advanced physics and mathematics. These are the very aspects that have shrouded the work of physicists in a cloud of esoteric mystery, discouraging the interest of the uninitiated. But this book is intended to show that the issues raised by the results of the LHC

---

[2] Resolution 4 of the 19th meeting of the Conférence générale des poids et mesures (1991).

are fascinating and of interest to anyone who believes it worthwhile to ask fundamental questions about nature.

Evidently the governments of the 20 European states members of CERN believe that it is worthwhile to ask these questions, for they have invested significant resources in the enterprise. The construction of the LHC accelerator amounted to about €3 billion, including the tests, the building of the machine, and the CERN contribution to the LHC computing and to the detectors, but without including the cost of CERN manpower. This enormous financial effort would not have been possible without substantial contributions from many countries that are not members of CERN: Canada, India, Japan, Russia, and the USA, among others. The design, construction, and testing of the instrumentation was done with the participation of physicists from 53 countries and five continents (regrettably no physicist or penguin from Antarctica decided to join the effort). The LHC is a stunning example of international collaboration in the name of science. As the LHC was built with the funding, labour, and intellectual contribution from so many different countries, its results are valuable property of all mankind. These results are not obtained just for the benefit of a few physicists, and their technical and specialist nature should not hide the importance of their universal intellectual content.

The LHC is the most complex and ambitious scientific project ever attempted by humanity. Every challenge met in its design and construction required advances in the frontiers of technology. The research that led to the LHC will certainly have spin-offs and practical applications beyond their purely scientific usage. After all, the World Wide Web was invented at CERN in 1989 to allow physicists to exchange data and information between laboratories in different parts of the globe. Four years later, CERN decided to release it into the public domain and, by doing so, gave to the world an instrument that has now become irreplaceable in our everyday life. Fundamental research often bears unexpected applications. In the middle of the 19th century William Gladstone, Chancellor of the Exchequer, asked the physicist Michael Faraday, engaged in research on electromagnetism, what could be the use of his discoveries. "I don't know, sir," replied Faraday, "but one day you will be able to tax it."

But for physicists, the ultimate goal of the LHC is only pure knowledge. Science enriches society well beyond any of its technological applications. In 1969 Robert Wilson, the director of a leading American laboratory for particle physics, was summoned to testify before Congress. The debate was about the possible justifications of spending $200 million on a particle physics project. Senator John Pastore, of the Congressional Joint Committee on Atomic Energy, questioned Wilson. In his reply, Wilson poignantly expressed the meaning of fundamental research.

PASTORE: Is there anything connected in the hopes of this accelerator that in
             any way involves the security of this country?
WILSON: No, sir. I do not believe so.
PASTORE: Nothing at all?
WILSON: Nothing at all.
PASTORE: It has no value in that respect?
WILSON: It only has to do with the respect with which we regard one another,
             the dignity of men, our love of culture. . . . It has nothing to do directly
             with defending our country, except to make it worth defending."[3]

This book deals with the journey made by the LHC: why it was undertaken and what we want to learn from it. This subject is inherently vast,
complicated, and of a highly technical nature, while the scope of this
book is comparatively modest. I will not cover all the topics systematically and I have no pretensions about giving a full account of the LHC
story. My aim is to give just a glimpse of the issues at stake from a
physicist's point of view, while underlining the intellectual broadness
and depth of the questions addressed by the LHC. I want to help the
reader to understand the meaning of this journey and why the whole
scientific community of particle physicists is so excited and is so eagerly
awaiting its results.

The first part of this book deals with the particle world and the way
physicists came to understand it. The results from the LHC cannot be
appreciated without some notion of what the particle world looks like.
As the theoretical physicist Richard Feynman once said: "I do not
understand why journalists and others want to know about the latest
discoveries in physics even when they know nothing about the earlier
discoveries that give meaning to the latest discoveries."[4]

The LHC is a machine of superlatives, where technological complexity
is pushed to the extreme. The second part of this book describes what
the LHC is and how it operates. The technological innovations required
to build the LHC were but one of the many astonishing aspects of this
scientific adventure. We will also encounter the detectors used to study
the particles created in the collisions between protons at the LHC. These
instruments are modern wonders that combine cutting-edge microtechnology with gigantic proportions.

The LHC is a project designed primarily for the exploration of the
unknown. So this book culminates with an outline of the scientific aims
and expectations at the LHC. The third part addresses some of the most

---

[3] Transcript of the Congressional Joint Committee on Atomic Energy (1969),
reprinted in *Proceedings of the American Philosophical Society*, vol. 146, no. 2, 229
(2002).

[4] R.P. Feynman, as quoted in S. Weinberg, *The Discovery of Subatomic Particles*,
Cambridge University Press, Cambridge 2003.

important questions about the goals of the LHC. How do physicists imagine zeptospace? Why should the mysterious Higgs boson exist? Does space hide supersymmetry or extend into extra dimensions? How can colliding protons at the LHC unlock the secrets of the origin of our universe? Is it possible to produce dark matter at the LHC?

Note to the reader

In particle physics, very large and very small numbers are often used. Therefore, in some cases, I will be forced to opt for scientific notation, although I will avoid it whenever possible. Thus, it is necessary to know that 10^{33} is equal to the digit *one* followed by 33 *zeros*: 1 000 000 000 000 000 000 000 000 000 000 000. I could also say *a million billion billion billion*, but 10^{33} looks a lot more concise and readable. Similarly, 10^{-33} is equal to 33 *zeros* followed by the digit *one*, with the decimal point after the first *zero*. So, 10^{-33} is *a millionth of a billionth of a billionth of a billionth*, or 0.000 000 000 000 000 000 000 000 000 000 001.

No previous knowledge of particle physics is required in order to read this book. I have limited as much as possible the use of technical terms but, when deemed unavoidable, their meanings have been defined in the text. For the ease of the reader, a glossary of technical terms has been included at the end of the book.

PART ONE
A MATTER OF PARTICLES

2
Dissecting Matter

It would be a poor thing to be an atom in a universe without physicists.

George Wald[1]

A droplet of oil on the surface of water cannot spread infinitely, but only up to a spot whose size is determined by the oil's molecular thickness. Salt dissolves in water only up to a maximum concentration, beyond which it sinks to the bottom of the container. These are simple indications for an empirical fact of nature: matter is not a continuous substance, but comes in lumps.

The apparently simple result that matter comes in lumps hides in reality some of the most astounding secrets of nature. Inside matter we discover unexpected new worlds, revolutionary fundamental principles, and unfamiliar phenomena that defy our intuition and contradict our sensory experience. But the most important result found in the depth of matter is that nature reveals a pattern. Hidden behind the complexity of our world lie simple fundamental laws that can be comprehended only by penetrating into the smallest constituents of matter. Dissecting matter is all about the quest for these fundamental laws of nature. As Richard Feynman put it: "If, in some cataclysm, all of scientific knowledge were to be destroyed, and only one sentence passed on to the next generation of creatures, what statement would contain the most information in the fewest words? I believe it is the atomic hypothesis.... In that one sentence, you will see, there is an enormous amount of information about the world, if just a little imagination and thinking are applied."[2]

[1] G. Wald, Preface to L.J. Henderson, *The Fitness of the Environment,* Beacon, Boston 1958.

[2] R.P. Feynman, *The Feynman Lectures on Physics*, Addison-Wesley, Reading 1964.

Atoms

> Democritus called it atoms. Leibniz called it monads. Fortunately, the two men never met, or there would have been a very dull argument.
>
> Woody Allen[3]

Leucippus and his disciple Democritus, philosophers active in Thrace between the fifth and the fourth century BC, claimed that matter is composed of atoms (from the Greek word *átomos*, indivisible) and of empty space. Aristoxenus narrates that Plato loathed the atomists' doctrine so much that he expressed the wish to burn all their writings existing in circulation. We do not know if Plato put this desire into action, but time certainly did. One single fragment from Leucippus and 160 fragments from Democritus are all that is left to us today, and very few of these make explicit reference to atoms. Most of what we know about the beliefs of the first atomists comes from later philosophers and historians.

According to Leucippus and Democritus matter is formed by a few species of fundamental atoms, differing in size and shape. Nature's complexity follows from the manifold combination of atoms and from their positions in empty space. Attributes of matter, like taste or temperature, are just the global effect of underlying microscopic entities. In other words, they are the consequence of a deeper structure in nature – that of atoms. Two of Democritus' fragments recite: "That in reality we do not know what kind of thing each thing is or is not has been shown many times. . . . By convention sweet and by convention bitter, by convention hot, by convention cold, by convention color; but in reality atoms and void."[4]

Tradition says that the ancient atomists were inspired by odours to formulate their ideas: matter is made up of atoms that can break away from substances and reach our noses. But the ancient concept of atoms is chiefly a philosophical assumption, more suited to address Zeno's difficulties with the idea of an infinitely divisible space, rather than able to explain specific observations of natural phenomena. Undoubtedly, some of the statements contained in the atomists' fragments strike us for their affinity with the modern vision, but of course their notions were very different from the reality that we understand today. For instance, tradition attributes to Democritus the idea that the different states of matter are associated with different atoms: round and smooth are the atoms of liquids, while those of solid substances have shapes fit to hook onto each other.

---

[3] W. Allen, *Getting Even,* First Vintage Books, New York 1978.

[4] C.C.W. Taylor, *The Atomists: Leucippus and Democritus; Fragments*, University of Toronto Press, Toronto 1999.

The atomists' view was a prophetic intuition that had no more empirical validity than the Aristotelian dogma according to which the fundamental elements (air, fire, earth, water) must be continuous entities. Atomism entered science only when it was invoked to explain the properties of gases, starting with the work of Isaac Newton, and to interpret the ratios between the different components in chemical reactions, starting with John Dalton. During the 19th century, many thermodynamical properties of gases began to be understood under the hypothesis that matter is not a continuous substance, but is made of some fundamental bits. This led to a new understanding of the structure of matter: all gases are composed of individual molecules. In turn, these molecules are compounds of some truly fundamental entities – the atoms.

In spite of the success of this hypothesis in explaining certain phenomena, some scientists were reluctant to accept the atomistic view. This was especially true in part of the German-speaking community, which was strongly influenced by the positivism of the Austrian physicist and philosopher Ernst Mach, who refuted the physical reality of entities – like the atoms – that could not be directly observed. This aversion to atomism contributed to the depressive state that brought about the suicide of Ludwig Boltzmann, the great Austrian physicist who fathered statistical mechanics.

Very different was the situation in England, where the tradition of Newton and Dalton favoured a consideration of atomism free from philosophical bias. It is therefore probably not simply fortuitous that the fundamental discoveries that established the reality of atoms took place in England. But, paradoxically, the incontrovertible evidence for the existence of the atom – the *indivisible* – occurred only when the atom was split.

Splitting the atom

It is harder to crack a prejudice than an atom.

Albert Einstein[5]

1897 is the year officially credited for the discovery of the electron, and the protagonist is Joseph John Thomson (1856–1940, Nobel Prize 1906). Thomson graduated from Cambridge University in 1880, and only four years later was elected Cavendish Professor, an appointment that provoked surprise in academic circles. Indeed, the chair, previously occupied by physicists of the calibre of Maxwell and Rayleigh, was one of the most prestigious in the world, and Thomson was just 28 years old

---

[5] Attributed to A. Einstein.

at the time. Moreover, it was a chair in experimental physics, while Thomson had, until then, been working mostly in theoretical physics and in mathematics. But the choice turned out to have been extremely foresighted.

Following his appointment, Thomson undertook the study of *cathode rays*. This is a form of radiation that is generated between two metal plates connected to a high-voltage power supply, when the apparatus is placed inside a glass tube, evacuated by pumping out all the air. Cathode rays were thought to be some form of electromagnetic radiation, although the French physicist Jean Baptiste Perrin (1870–1942, Nobel Prize 1926) had found the perplexing result that these rays appeared to deposit electric charge on the metal plate. If true, this was in contradiction with the initial hypothesis, since electromagnetic radiation carries no electric charge.

Thomson addressed the problem by applying an electric field inside the glass tube in order to understand if it could influence the cathode rays. He observed a deflection in the trajectory of the cathode rays. This was irrefutable evidence that the rays carry electric charge and cannot be a form of electromagnetic radiation. Others before him had tried this experiment, but failed to detect any measurable effect. Thomson succeeded, thanks mostly to more powerful vacuum pumps that allowed him to reduce the pressure of the residual gas inside the tube.

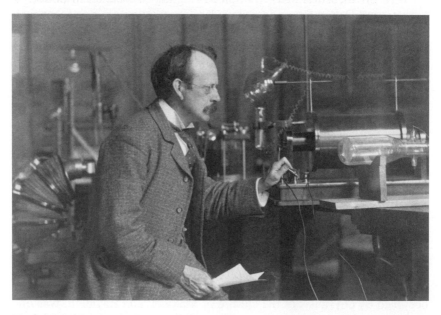

Figure 2.1 Joseph John Thomson giving a lecture demonstration at the University of Cambridge.

Source: Cavendish Laboratory / University of Cambridge.

The apparatus used by Thomson is nothing other than a primitive version of the cathode tube used in old-fashioned television sets. Just as in Thomson's experiment, appropriate electric fields inside a television continually deflect the beam of cathode rays which, hitting a fluorescent screen, leaves a luminous dot. These dots quickly change their positions on the television screen, but our eyes' retinas react slower, overlapping the images and thus creating the effect of the full picture that is perceived by our brain.

Thomson repeated his experiment by applying various electric and magnetic fields to his apparatus, each time measuring the deflection of the trajectory of the cathode rays. Once he collected the data, he drew his conclusions. He started with the hypothesis that the cathode rays were made of electrically charged particles and he computed the deflection of the beam subjected to electric or magnetic forces. Comparing the theoretical calculation with his measurements, he was able to deduce the ratio between mass and electric charge of the hypothetical particles. He found that this ratio was about a thousand times smaller than for a hydrogen ion – the lightest known chemical element. Thomson had no doubts and boldly concluded: "On this view we have in the cathode rays matter in a new state, a state in which the subdivision of matter is carried very much further than in the ordinary gaseous state."[6] In other words, the atoms had been split, and one of its fragments had been observed.

With his measurements Thomson had only succeeded in deducing the ratio between mass and electric charge of the atom's fragment, and not the two quantities separately. There was still something to understand: "The smallness of m/e [the ratio between mass and charge] may be due to the smallness of m [the particle mass] or the largeness of e [the particle charge] or to a combination of both."[7] Two years later Thomson managed to make a first rough measurement of the electric charge, later refined by Robert Millikan (1868–1953, Nobel Prize 1923) and his student Harvey Fletcher (1884–1981). This was the confirmation that the fragment was much lighter than the full atom. The *electron* had been discovered.

This discovery opened a new chapter in physics, because it demonstrated that the atom could be split. Moreover, Thomson had identified the substance that carries the electric charge in a flow of current. The conclusion was that electrical phenomena are caused by the breakaway of electrons from atoms, as Thomson himself explained: "Electrification essentially involves the splitting up of the atom, a part of the mass of the atom getting free and becoming detached from the original atom."[8]

---

[6] J.J. Thomson, *Philosophical Magazine* 44, 295 (1897).

[7] J.J. Thomson, *ibid.*

[8] J.J. Thomson, *Philosophical Magazine* 48, 547 (1899).

It was certainly a bit of a gamble for Thomson to conclude, on the basis of the 1897 results, that he had discovered a "new state of matter" and that he had identified an atom's fragment – the electron. After all, what he had actually observed was just a displacement of cathode rays; the rest was rather speculative deduction. In 1897, a few months before Thomson completed his study, Walter Kaufmann (1871–1947) in Berlin had obtained and published very similar experimental results. Kaufmann had measured deflections in cathode rays and noticed their independence of the kind of residual gas inside the glass tube. The smallness of the mass-to-charge ratio deduced from the experiment appeared to him so absurd as to make him conclude that the hypothesis of a particle nature of cathode rays must be wrong: "I believe to be justified in concluding that the hypothesis of cathode rays as emitted particles is by itself inadequate for a satisfactory explanation of the regularities I have observed."[9] In summary, he obtained the same experimental results as Thomson, but drew opposite conclusions.

Thomson is credited with the discovery of the electron, while Kaufmann's work is completely ignored by physics textbooks. Certainly the scientific atmosphere of Berlin University, reluctant to entertain any corpuscular interpretation, played against Kaufmann. But in physics, merit goes to those who have the intuition to see in a phenomenon the interpretative key to unlock the secrets of nature, and Thomson had this intuition. As the Nobel Prize laureate physiologist Albert Szent-Györgyi elegantly put it: "Discovery consists of seeing what everybody has seen and thinking what nobody has thought."[10]

Inside the atom

If this is true, it is far more important than your War.

Ernest Rutherford (message sent during World War I to a military research committee to justify his absence while engaged in his experiments on the atomic nucleus)[11]

Once the reality of the electron had been established, it remained to discover what was the substance that comprises the rest of the atom and neutralizes its total electric charge. Some fragments of atoms – the electrons – had been observed, but these fragments constituted only a tiny fraction of the total atomic mass. What was the rest made of?

[9] W. Kaufmann, *Annalen der Physik und Chemie* 61, 544 (1897).

[10] A. Szent-Györgyi, as quoted in *The Scientist Speculates*, ed. I.J. Good, Heinemann, London 1962.

[11] E. Rutherford, as quoted in T.E. Murray, *More Important Than War*, in *Science* 119, 3A (1954).

Thomson conceived the atom as a uniform entity of positive electric charge, with electrons trapped inside. This picture was known as "Thomson's plum pudding" because the electrons looked like specks of dried fruit inside a sticky substance. The model – let alone this very Anglo-Saxon dessert – wasn't very enticing to non-British palates. Indeed, in 1903 the Japanese physicist Hantaro Nagaoka (1865–1950) proposed the idea of an atom similar to the solar system, with a "sun" in the middle and the electrons interpreted as "planets" orbiting around. Hermann Helmoltz and Jean Baptiste Perrin considered a similar idea too. However, the hypothesis of an atomic "solar system" was untenable. In fact, it was well known that an orbiting electric charge emits electromagnetic radiation, losing energy, and therefore the electrons would rapidly fall into the centre, making the atom quickly collapse. The atomic structure remained a mystery.

Ernest Rutherford (1871–1937, Nobel Prize 1908) was a brilliant student from New Zealand who, thanks to a grant, moved to the glorious Cavendish Laboratory in Cambridge, full of hopes and ambitions. Later in his life, he became a physics professor at the University of Manchester. One day in 1909, in Manchester, he suggested to his collaborator Hans Geiger (1882–1945) and to his student Ernest Marsden (1889–1970) to study the diffusion of the so-called *alpha particles* (which are positively charged helium ions) produced by a radioactive source of radium bromide. Diffusion occurs when the alpha particles hit a thin film of gold or aluminium and, while passing through, their original trajectories are modified. Experiments of this kind had already been performed and it was observed that the alpha particles are slightly deflected when they cross the film. The novelty of Rutherford's suggestion was that he asked his collaborators to check if any alpha particle bounced back instead of going through the film.

The project proposed by Rutherford to his collaborators sounds suspiciously like one of those problems that you give to students just to keep them busy, before you can think of a better idea to work on. Why on earth should a thin metal film reflect heavy and fast-moving bullets, like the alpha particles produced by a radioactive source?

Geiger and Marsden made their measurement and ran back breathlessly to Rutherford. They had observed that some alpha particles were indeed bouncing back. In Rutherford's words: "It was quite the most incredible event that has ever happened to me in my life. It was almost as incredible as if you fired a 15-inch shell at a piece of tissue paper and it came back and hit you."[12] For those fast and heavy alpha particles to come bouncing back, they would have had to meet inside the film an

[12] Quoted in E.N. da Costa Andrade, *Rutherford and the Nature of the Atom*, Doubleday, New York 1964.

Figure 2.2 Ernest Rutherford (right) and Hans Geiger at the Schuster Laboratory of the University of Manchester.

Source: Bettmann Archive/Corbis/Specter.

obstacle and a force so intense as to completely reverse their motion. This obstacle could not be provided by the electrons, which are too light. It is like violently throwing a bowling ball against few ping-pong balls. You cannot expect the bowling ball to bounce backwards. Neither was the sticky substance of Thomson's plum pudding adequate to reflect the energetic alpha particles.

Throughout his life, Rutherford had been an inveterate and ingenious experimental physicist, but he was always rather sceptical of the activity of most theoretical physicists, which he considered too speculative and abstract. Yet, just this once, he played the game of theoretical physics. He computed the probability that an alpha particle could be deflected by an angle larger than 90 degrees (in other words, that it could be reflected) under the hypothesis that the whole mass and positive electric charge of the atom be concentrated in a single point – an *atomic nucleus*. The result of the calculation turned out to be in perfect agreement with the data found by Geiger and Marsden, and also with later experiments performed by Rutherford in collaboration with Marsden.

The alpha particles, which have a positive electric charge, penetrate into the metal film and, in general, their trajectories suffer only minor disturbances caused by the electromagnetic forces exerted by the various atomic charges within the foil. This explains the small deflections observed in the diffusion of the majority of alpha particles. However, there is a probability, though very small, that the trajectory of an alpha particle comes very close to a nucleus, where all the atomic mass and positive electric charge are concentrated. In this case, the alpha particle can be reflected because the electromagnetic force in the vicinity of the heavy nucleus is very intense. It is like throwing a bowling ball against a stationary and large cannon ball. This time there is a chance that the bowling ball will bounce backwards. Actually the same thing occurs in the deflection of a comet's trajectory. When a comet goes through an asteroid belt, its trajectory is hardly modified, but when the comet approaches the sun, it is affected by the intense force of gravity and can be deflected by a large angle along a hyperbolic trajectory.

Rutherford had looked inside the atom and the image he saw was very different from what physicists had expected. A central nucleus, much smaller than the actual size of the atom, holds the entire positive charge and practically all the atomic mass. The rest is just a cloud of light electrons, carrying all the negative charge.

Just for curiosity we can compare sizes and weights of the solar system with those of the atom. The ratio between the size of the solar system (choosing Neptune's orbit as its border) and the diameter of the sun is about 6000:1, and the ratio between the mass of the sun and those of all the planets is about 700:1. For intermediate-size atoms, the size ratio between the atom and the nucleus is about 20 000:1, and the mass ratio between nucleus and electrons is almost 4000:1. Therefore, in comparison, the atom is much more empty than the solar system and its mass is much more concentrated in the centre. Rescaling proportions, the dimension of a nucleus inside the atom is that of "a fly in a cathedral."[13]

Nonetheless, as previously noted, the picture of atoms as miniature solar systems was totally inconceivable according to the laws of electromagnetism. Then the Danish theoretical physicist Niels Bohr (1885–1962, Nobel Prize 1922) made the hypothesis that the electrons inside the atom must occupy only special orbits. In the case of the solar system, the distances between the sun and the planets are not determined by any fundamental principle. No physical law forbids the existence of other solar systems where the distances between the central star and the orbiting planets are different from those in our system. According to Bohr, this is not true for electrons. Only certain special orbits are possible; anything else is forbidden.

Imagine a tourist in Egypt who wants to take a picture with a nice panoramic view of the desert. To have a better viewpoint, he needs to go

---

[13] J. Rowland, *Understanding the Atom*, Gollancz, London 1938; B. Cathcart, *The Fly in the Cathedral*, Viking, London 2004.

to an elevated place, but there are no hills in the area. Suddenly he has the bright idea of climbing up the sides of the Great Pyramid of Giza, which we imagine as perfectly smooth. He can choose at will the height from which to take his picture, just by climbing a little further up or descending a bit. A few days later, the same tourist goes to Saqqara to visit the famous Step Pyramid. Here he feels the same urge to take a picture from an elevated place. He starts climbing the Saqqara pyramid, but this time only certain heights are accessible – those determined by the steps of the pyramid. Every intermediate height is excluded, because the tourist would immediately slide onto the level of the lower step. In the same way, while planets in a solar system can occupy any orbit, only special well-defined levels are accessible to electrons inside the atom.

Starting from this hypothesis, Bohr invented new rules for the motion of particles. These rules would lead to the birth of a new theory: *quantum mechanics*. This new theory would soon undermine the Newtonian description of motion and revolutionize many fundamental physics concepts. Even the notions of trajectory and of orbit lose their ordinary meaning in quantum mechanics.

Bohr's simple, but strange, hypothesis of the electrons' orbits was not initially justified by any sensible physics principle. Nevertheless, not only was it adequate to explain atomic structure, but it could also predict the *frequency spectrum* of hydrogen. The spectrum of a chemical element is the set of frequencies of light that are absorbed or emitted by that element. These frequencies are a distinctive feature of a given

Figure 2.3 Niels Bohr (right) discussing with Werner Heisenberg.
Source: Pauli Archive / CERN.

chemical element and they provide its fingerprints with which it can be identified uniquely. For instance, sodium lamps do not emit white light – that is, distributed among all frequencies – but only light with two special values of frequency. Since the frequency of light is what we perceive as colour, sodium lamps appear to our eyes with the distinctive yellow-orange light often visible on certain motorways.

Conversely, by decomposing with a prism white light that has gone through some gas, one discovers black lines that exactly coincide with the characteristic frequencies, or the fingerprints, of that gas. The element that composes the gas has absorbed the frequencies of light corresponding to its spectrum. The analysis of the frequency spectrum of light coming from stars led to a fundamental discovery of science in the 19th century: the chemical elements present in the celestial bodies are exactly the same as those existing on earth – stellar elements show identical fingerprints to terrestrial elements. Curiously, the chemical element helium was first discovered in the sun and only later on earth, as its name reminds us (from the Greek *helios*, sun).

Bohr assumed that the frequency spectrum of an element corresponds to the energy differences between the possible electron orbits inside the atom. In the previous analogy, the frequency corresponds to the energy required to jump from one step of the Saqqara pyramid to another. Since there are only a limited number of possible steps, the frequencies of an element's spectrum are given by a few special values. Bohr could then compute the frequency spectrum of hydrogen, which was already known experimentally with great precision. The agreement between Bohr's result and measurements was absolutely astonishing.

The discovery of the atomic nucleus had not only disclosed the most intimate structure of matter, but had also revealed that the fundamental laws of nature describe a world that is very different from that which is usually perceived by our senses. The weirdness of Bohr's hypothesis and its success in explaining the properties of the hydrogen atom left many in absolute bewilderment. It is said that, in those days, the most common question among theoretical physicists was: "Do you believe it?" Bohr himself probably gave the most adequate answer to this question, although in a completely different context. One day a guest went to visit Bohr in his country house at Tisvilde, Denmark, and he was surprised to find a horseshoe hanging over the front door. He asked Bohr if he really believed that a horseshoe could bring good luck. "Of course not," replied Bohr, "but I am told that it works even if you don't believe in it."[14] The same thing could have been said for the first hypotheses of quantum mechanics. Nobody had any good rational justification

[14] P. Robertson, *The Early Years, the Niels Bohr Institute 1921–30*, Akademisk Forlag, Copenhagen 1979.

of why the strange rules of quantum mechanics worked, but nonetheless they could triumphantly explain the experimental observations made on atoms and on their internal structures. But the surprises of quantum mechanics had just started and much more was to come.

Inside the atomic nucleus

> Nuclear powered vacuum cleaners will probably be a reality within ten years.
>
> Alex Lewyt, president of the Lewyt Vacuum Cleaner Company, interviewed in 1955[15]

Rutherford's discovery of the atomic nucleus and Bohr's theory of electron orbits paved the way for measurements of the positive electric charge contained in the atoms. The idea was to shoot an X-ray beam onto atoms in order to hit some electrons, ejecting them out of their orbits. The remaining electrons would reorganize their structure, filling some empty orbits of lower energy, thus emitting secondary X-rays that could be measured. The frequencies of the secondary X-rays carried the information about the energy levels in inner electron orbits. Through theoretical calculations and data on X-ray frequencies, it was possible to deduce the electric charge of the nucleus, the so-called *atomic number* Z.

Systematic measurements of the atomic number of almost all the known elements were carried out by Henry Moseley (1887–1915), who had developed at Oxford new techniques for the determination of X-ray frequencies. His brilliant career was prematurely cut short. When the First World War broke out, the 26-year-old Moseley volunteered for the British Army and fell at Gallipoli under Turkish fire. In the meanwhile, the British Marsden and the German Geiger, the two collaborators of Rutherford, were both fighting at the Western Front, but on opposite sides.

By measuring atomic numbers, physics was rediscovering Mendeleev's periodic table and was finding a new and deeper meaning to the classification of chemical elements. The values of the atomic number Z (the nucleus electric charge) were found to be integers, at least within the experimental error. Moreover, the heavier the element was, the larger the value of Z. The weight of an element in units of the hydrogen atom is called the *atomic weight* A. The values of A for the various elements were known from chemistry to be approximately integer numbers as well.

All these were very good clues to the idea that nuclei are compounds of simpler entities, of which the hydrogen nucleus is the fundamental

---

[15] *Vacuum Cleaners Eyeing the Atom*, The New York Times, 11 June 1955.

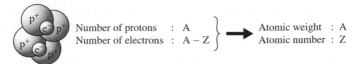

Figure 2.4 The nucleus of helium according to the nuclear model with A protons and A − Z electrons. Helium has atomic weight A = 4 and atomic number Z = 2.

building block. This building block has a positive electric charge, equal and opposite to that of the electron, but a mass 1836 times larger than the electron mass. Starting in 1920, the hydrogen nucleus – the building block of all nuclei – was called the *proton* (from the Greek *protos*, first), a name first used by Rutherford.[16]

However, it was immediately clear that atomic nuclei could not be made of only protons. If protons were the only constituents, then the total nuclear mass and charge would be the sum of the masses and charges of individual protons in the nucleus. Thus each chemical element should have equal values of A and Z, since A and Z count the nuclear mass and charge in proton units, respectively. Measurements contradicted this expectation: the value of A was growing from element to element faster than Z. For instance, Moseley had found Z = 22, A = 48 for titanium, Z = 23, A = 51 for vanadium, Z = 24, A = 52 for chromium, and so on.

At that time, the accepted hypothesis to explain these observations was that the nucleus contained protons and electrons. Since the mass of the electron is virtually negligible with respect to the proton, the number of protons in the nucleus must be equal to A. Moreover, as illustrated in Figure 2.4, the nucleus had to contain a number of electrons equal to A − Z, because the atomic charge is given by the number of protons minus the number of electrons (electrons carry negative charge).

Today we know that this explanation is wrong but, at that time, it was certainly the most plausible alternative. Electrons and protons were the only known particles and therefore the natural ingredients for any atomic recipe. Moreover, it was known that beta-radioactive nuclei emit electrons, and therefore it was perfectly reasonable to believe that electrons existed inside the nucleus. Finally, it was not clear what could keep protons together inside the nucleus. After all, electric forces between positive charges are repulsive, and should quickly disintegrate the nucleus. Although nobody was able to propose a credible explanation of nucleus stability, at least adding electrons was giving some hope. This is because electrons have a negative charge and would therefore exert an attractive force on protons. In Rutherford's words: "The nucleus, though

[16] *Physics at the British Association,* in *Nature* 106, 357 (1920).

of minute dimensions, is in itself a very complex system consisting of positively and negatively charged bodies bound closely together by intense electrical forces."[17]

And yet, some theoretical physicists were objecting to the idea of nuclei composed of protons and electrons. Initial knowledge of quantum mechanics indicated that it is not possible to confine electrons in spaces as small as the atomic nuclei. For their part, experimental physicists were trying to bombard protons with electrons in the attempt to neutralize the total electric charge and to form those "very complex systems" that were believed to exist in nuclei. These attempts failed year after year.

Then events unfolded rapidly. On 28 January 1932, Irène Joliot-Curie (1897–1956, Nobel Prize 1935) and Frédéric Joliot-Curie (1900–1958, Nobel Prize 1935) – the daughter and the son-in-law of the famous Curie couple – announced their discovery: beryllium atoms, after being bombarded with alpha particles, emitted certain rays able to extract protons out of a target made of paraffin wax. The Joliot-Curies erroneously interpreted these rays as electromagnetic gamma radiation. The puzzling aspect of their result was that, for this electromagnetic radiation to be penetrating enough to extract protons from paraffin, it would need an energy much larger than was available inside a beryllium atom. The Joliot-Curies even speculated on the possibility that the energy, at the nuclear level, is not conserved.

When Ettore Majorana (1906–1938?), the Italian physicist who later mysteriously disappeared without leaving a trace, first heard about the result, he exclaimed: "Oh, look at the idiots; they have discovered the neutral proton, and they don't even recognize it."[18] Sicilians, like Majorana, are known to be less gracious (but possibly more expressive) than Englishmen. Indeed James Chadwick (1891–1974, Nobel Prize 1935) reacted to the same event with a more restrained comment: "An electrifying result."[19]

Chadwick, as well as Majorana, had immediately understood that the radiation, to be so penetrating, had to be caused by a neutral and heavy particle. The publication by Joliot-Curie had reached Cambridge at the beginning of February. Chadwick worked in his laboratory for ten straight days, without abandoning his other responsibilities at the Cavendish, sleeping no more than three hours a night. On 17 February 1932, he sent to the scientific journal *Nature* his article describing the discovery

---

[17] E. Rutherford, *Scientia* 16, 337 (1914).

[18] As recalled by Gian Carlo Wick and by Emilio Segrè. See A. Martin in *Spin in Physics*, ed. M. Anselmino, F. Mila and J. Soffer, Frontier, Turin 2002, and E. Segrè, *Nuclear Physics in Retrospect: Proceedings of a Symposium on the 1930s*, ed. R.H. Stuewer, University of Minnesota Press, Minneapolis 1979.

[19] J. Chadwick, in *Proceedings of 10th International Congress on the History of Science*, Ithaca, New York, Hermann, Paris 1964.

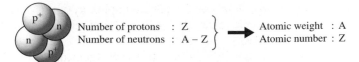

Figure 2.5 The nucleus of helium according to the nuclear model with Z protons and A – Z neutrons. Helium has atomic weight A = 4 and atomic number Z = 2.

of the *neutron*. Soon after, he gave a seminar at the Cavendish where he explained to his colleagues his findings and concluded with the words: "Now I want to be chloroformed and put to bed for a fortnight."[20]

Chadwick had discovered the neutron, a particle with zero electric charge and with a mass almost identical to the proton mass. Actually, both Chadwick and Rutherford still believed that the neutron was a bound state of a proton and an electron. Later theoretical studies clarified that the neutron is, in all respects, a particle and an ingredient of the nucleus, as much as the proton.

So the picture of the atom changed. The nucleus is composed by a number Z of protons and a number A – Z of neutrons (see Figure 2.5). This correctly explains its total mass and electric charge. The electrons occupy only orbits external to the nucleus and fill most of the space inside the atom.

The discovery of the neutron had unexpected consequences. The Hungarian physicist Leo Szilard (1898–1964) recounts: "I remember very clearly that the first thought that liberation of atomic energy might in fact be possible came to me in October 1933, as I waited for the change of a traffic light in Southampton Row in London.... It occurred to me that neutrons, in contrast to alpha particles, do not ionize the substance through which they pass. Consequently, neutrons need not stop until they hit a nucleus with which they may react."[21] Szilard had realized that neutrons, having no electric charge, are not stopped by the electromagnetic barrier that protects the nucleus. Then, even a relatively slow neutron can penetrate into a nucleus and possibly split it, freeing some of large energy stored within. The break-up of the nucleus could produce the leak of other neutrons that, in turn, would split more nuclei in a chain reaction. Nature had put into the hands of physicists an arrow – the neutron – able to strike the most intimate part of matter.

When the traffic light in Southampton Row turned green, a new adventure for humanity had started. It was an adventure that had to do with war and horror. But that is a different story from the one I am telling here.

[20] C.P. Snow, *The Physicists*, Little Brown, Boston 1981.

[21] L. Szilard, *The Collected Works: Scientific Papers*, ed. B.T. Feld and G. Weiss Szilard, MIT Press, Boston 1972.

Figure 2.6 Paul Dirac (right) with Wolfgang Pauli in Oxford in 1938.
Source: Pauli Archive / CERN.

Antimatter

> Communication is established between humans here on earth and extraterrestrials living in a galaxy made of antimatter. It is found that in that anti-world they have anti-science, anti-mathematics, and anti-physics. Earthbound physicists get a description of an anti-physics anti-laboratory, and lo and behold, they find it is filled with anti-Semites.
>
> Peter Freund[22]

Sometimes progress in physics is driven by important experimental discoveries; sometimes it follows from new theoretical speculations; more often it is a combination of both. Antimatter is an example of a concept generated primarily by pure thought, and only later confirmed by experiments. The logical path that led to the discovery of antimatter was tortuous and difficult. Few people could have blazed this trail better than Paul Dirac (1902–1984, Nobel Prize 1933).

Dirac was born in Bristol of a Swiss father, of whom he recalls: "My father made the rule that I should only talk to him in French. He thought

---

[22] P. Freund, *A Passion for Discovery*, World Scientific, Singapore 2007.

it would be good for me to learn French in that way. Since I found that I couldn't express myself in French, it was better for me to stay silent than to talk in English. So I became very silent at that time – that started very early."[23] Dirac's discretion was legendary. About his meetings with Bohr, he recalls: "We had long talks together, long talks in which Bohr did practically all the talking."[24] Dirac used words sparingly, but he could talk with equations.

At the beginning of the 20th century, special relativity and quantum mechanics revolutionized the world of physics. Special relativity reformulated the premises on how different observers in uniform relative motion perceive time intervals and distances in space. This theory showed that many basic concepts of classical physics are no longer valid for bodies with velocities close to the speed of light. On the other hand, quantum mechanics redefined our understanding of processes with small exchanges of energy. The two theories had revealed new realities, but kept separate domains. However, to describe the motion of electrons at high energies it was necessary to formulate a combined theory that could include the effects of both quantum mechanics and relativity. The construction of such a theory was presenting great difficulties from the mathematical point of view. But Dirac loved difficult problems and he was determined to find an equation that could exactly describe the electron's motion.

In 1928 Dirac obtained the equation he was looking for. I suppose that very few equations have the privilege of being displayed in a cathedral. Dirac's equation is engraved in stone in Westminster Abbey, adjacent to Newton's grave. This is already a remarkable achievement for an equation, but there is more. Dirac's equation not only gives a unified description of special relativity and quantum mechanics, but it also determines the magnetic properties of the electron, in perfect agreement with experiments.

The taciturn Dirac had discovered the common mathematical language with which special relativity and quantum mechanics could finally discourse. However, something wasn't quite right. Dirac's equation had a double solution: besides the electron, the equation was also describing some other mysterious entity, possibly another particle. This mysterious particle had the same mass as the electron and an electric charge equal in size, but positive. To make things worse, this particle had negative energy. What was the meaning of all this?

The confusion lasted more than three years. A particle with negative energy was seen as a catastrophe. A physical system always evolves towards the state of minimum energy. But, if a particle contributes a

---

[23] A. Pais, *Inward Bound*, Oxford University Press, Oxford 1986.
[24] P.A.M. Dirac, in *History of Twentieth Century Physics*, Academic Press, New York 1977.

Figure 2.7 The slate plaque engraved with Dirac's equation in the floor of Westminster Abbey.

Source: Westminster Abbey.

negative amount to the total energy of the system, then any increase in the number of such particles makes the total energy decrease. Therefore the universe should collapse into a gigantic clump of negative-energy particles.

To bypass this absurd result, Dirac made the hypothesis that the particles with negative energy should just be discarded from the solution of his equation, because they are not physical. But this cheap way out did not work. It was soon proved that the existence of the negative-energy particles was essential for the consistency of the theory. Without them, Dirac's equation could not reproduce the known quantum-mechanical result, in the limit of small electron velocity.

Then Dirac tried a different explanation. He made the hypothesis that these new positively charged particles were the protons, hoping that electromagnetic effects could justify the difference between the proton and electron masses. Even worse: in this case all atoms would disintegrate into gamma rays after only 0.1 nanoseconds.

Finally, in 1931, Dirac made the decisive step: the new particle, "if there were one, would be a new kind of particle, unknown to experimental physics, having the same mass and opposite charge of the electron."[25] Dirac called this hypothetical particle the anti-electron. At that time, Dirac had an explanation for the negative values of the energy, which today however is viewed as obsolete and not satisfactory. The complete understanding of the physical meaning of negative energy came only later, with developments in quantum field theory.

About a year later, the American physicist Carl Anderson (1905–1991, Nobel Prize 1936) discovered in his apparatus tracks of unusual particles: they had positive charge, but were much lighter than protons. He concluded that he had identified a new kind of particle with charge

---

[25] P.A.M. Dirac, *Proceedings of the Royal Society*, A133, 60, 1931.

opposite to that of the electron and of comparable mass. Anderson, unaware that the particle already had a name among theorists, called it the *positron*. This is the name most frequently used today. "Yes, I knew about the Dirac theory," declared Anderson in a later interview. "... but I was not familiar in detail with Dirac's work. I was too busy operating this piece of equipment to have the time to read his papers."[26]

So, Dirac had got it right after all although, as he later admitted: "The equation was smarter than I was."[27] The marriage between special relativity and quantum mechanics had generated antimatter. In other words, the logical consistency of special relativity and quantum mechanics implies that matter cannot exist without its counterpart – antimatter. For each particle there exists a corresponding antiparticle, a sort of mirror image of the particle. They both have exactly the same mass but opposite electric charges.

Anderson's discovery of the positron opened the hunt for antimatter. The next goal was to prove that protons and neutrons have their corresponding antiparticles too. To produce antiprotons and antineutrons, which are almost 2000 times heavier than positrons, high-energy particle accelerators were needed. Physicists at Berkeley started the construction of the Bevatron, an accelerator that generated a proton beam that could be used to bombard material targets. The proton energy was enormous for those days, but is actually less than a thousandth of the energy of a single LHC beam. In 1955, experiments at the Bevatron, led by Emilio Segrè (1905–1989, Nobel Prize 1959) and Owen Chamberlain (1920–2006, Nobel Prize 1959) discovered the antiproton. The following year it was the antineutron's turn.

Do anti-atoms exist? As atoms are composed of protons, neutrons and electrons, the combination of antiprotons, antineutrons and positrons can form anti-atoms. Although stable anti-atoms are not found in nature, they can be produced in a laboratory. The simplest form of anti-atom was first created at CERN in 1995. It is the anti-hydrogen atom, which is made of a single positron orbiting around one antiproton. The experiment was performed at the Low-Energy Antiproton Ring (LEAR), a 78-metre circumference ring, used to decelerate antiprotons, before combining them with positrons. After the shutdown of this machine, new experiments were carried out in an apparatus dedicated to anti-hydrogen production, the Antiproton Decelerator (AD). In 2002, several events corresponding to anti-hydrogen production and decay were recorded at CERN. The biggest technological challenge that remains is to trap, by means of magnetic fields, the anti-atoms for a time long enough to make precise measurements of their structure. Research in this direction is still in progress.

---

[26] A. Pais, *Inward Bound*, Oxford University Press, Oxford 1986.
[27] G. Johnson, *Strange Beauty*, Knopf, New York 2000.

In spite of the fact that research on artificial production of anti-atoms has had, at the moment, little impact on the scientific world, it has aroused great interest in the public and the media. Antimatter has even been invoked as a viable energy source. Unfortunately, the efficiency for producing energy from antimatter is ridiculously small and "all anti-matter ever produced at CERN would not even be enough to light a 100 W electric light bulb for more than one hour."[28] Still antimatter remains one of the favourite subjects of science-fiction writers.

[28] R. Landua, *Physics Reports* 403–404, 323 (2004).

3
Forces of Nature

When force is necessary, it must be applied boldly, decisively, and completely.

Leon Trotsky[1]

Some of the ancient thinkers had the intuition that all forms of matter could be ultimately ascribed to a few fundamental elements. Modern science has proved them right. But it would have been difficult for ancient philosophers to guess that this is true not only for matter, but for force as well. Less intuitive is the idea that all natural phenomena, in all their variety and complexity, can be reduced to four fundamental forces: gravity, electromagnetism, weak and strong interactions. Even more unforeseeable is the result that forces, like matter, are produced by elementary particles. The intellectual trail that led to this understanding was no easy ride.

Force of gravity

Gravity is a kind of mystical behaviour in the body, invented to conceal the defects of the mind.

François de La Rochefoucauld[2]

Aristotle explained gravity as a natural tendency of motion. Each body, not subjected to forces or external agents, follows its *natural* motion, defined by straight lines, upwards for light elements (air and fire) and downwards for heavy ones (earth and water). The heavier the body, the faster it falls. Similarly, it is in the nature of the earth to seek the centre

---

[1] L. Trotsky, *What Next? – Vital Questions for the German Proletariat,* 1932.

[2] F. de La Rochefoucauld, *Maximes*; reprinted in English in *La Rochefoucauld, The Moral Maxims and Reflections (1665-1678)*, ed. G. Powell, Stokes Company, New York 1930.

of the universe, and in the nature of celestial bodies to follow circular paths around it.

Aristotle maintained an *organicist* view, which attributed to inanimate bodies a natural tendency towards a global organization, almost in imitation of a human population. Starting around the 17th century, this doctrine was replaced by a *mechanistic* view in which physical laws, expressed in terms of mathematical equations, determine motion.

In this new approach to science, the use of experiments played a crucial role. One should not believe that Aristotle's view of the world was just the result of pure philosophy, for he had always maintained that observation of nature should be the starting point of any assertion. But there is an important difference between observation and experiment. In observation, natural phenomena are studied as they are presented to our senses. In experiment, one creates special situations, under controlled and reproducible conditions, to obtain quantitative information on nature's behaviour.

The merit for this change of attitude in science goes primarily to Galileo Galilei (1564–1642). From his studies, Galileo concluded that gravity accelerates all bodies in the same way, whatever their masses. This result marks a clear departure from Aristotelian doctrine. Experiments were at the origin of Galileo's assertion, but he had to extrapolate his data to a situation in which he could neglect the effect of air friction. Simple observations, in which the effects of gravity and friction are not separated, can lead to wrong conclusions.

We are full of admiration, imagining Galileo who, with his legendary arrogance and swaggering, quickly climbs up the spiral staircase of the Tower of Pisa. There, with a confident and mocking smile, he drops a heavy cannon ball and a light musket pellet out of the protruding side of the leaning tower. The two objects hit with perfect simultaneity the underlying meadow, greeted by the rejoicing of the crowd and by the fainting of some old university sages.

Alas, this story is certainly false. With today's knowledge, it was proved that Galileo could not have made this kind of public demonstration[3]. Because of air resistance, it would not work. Also, the typical human reaction time could not have allowed Galileo to drop the balls with the required simultaneity. Actually, there is no mention of this demonstration in any of Galileo's writings. The story comes from a biography by Vincenzo Viviani (1622–1703), Galileo's last assistant, who probably wanted to add some extra glory to the celebrated life of his master. The truth is that the law of gravitational acceleration is the

---

[3] G. Feinberg, *American Journal of Physics* 33, 501 (1965); B.M. Casper, *American Journal of Physics* 45, 325 (1977); C.G. Adler and B.L. Coulter, *American Journal of Physics* 46, 199 (1978).

result of careful and precise experiments on inclined planes, performed in the silence of a laboratory. So much the better.

Isaac Newton (1643–1727)[4] discovered the universal law of gravitation, and it took more than a falling apple to crack the problem. Newton computed the acceleration necessary to keep the moon in a stable orbit around the earth. Next, he noticed that the value he obtained was smaller than the gravitational acceleration on earth by a quantity equal to the square of the ratio of the earth–moon distance to the earth radius. The real breakthrough was to show that a gravitational force decreasing with the square of the distance leads to elliptical planetary orbits, with the sun situated at one of the foci. This is exactly the result of Kepler's law. Therefore the empirical law, derived by Kepler on the basis of astronomical observations, could be deduced from Newton's theory of gravity.

The crucial conceptual step made by Newton was to understand the universal quality of gravitation. The same force that makes apples fall from trees governs planetary motion. The same mathematical equation rules over completely different phenomena and allows us to compute the motion of bodies at any place in the universe.

More than 300 years later, Albert Einstein (1879–1955, Nobel Prize 1921) was disturbed by one aspect of the Newtonian theory of gravity. How does the earth know that the sun exists, 150 million kilometres away, and move accordingly? Newton's theory has no answer to this question. The force of gravity is conveyed instantaneously even at cosmic distances. But nothing in the theory explains how a body can act at a distance or how the force is transmitted around space. Newton himself was well aware of this limitation of his theory when he wrote: "That one body may act upon another at a distance through a vacuum without the mediation of anything else, by and through which their action and force may be conveyed from one another, is to me so great an absurdity that, I believe, no man who has in philosophic matters a competent faculty of thinking could ever fall into it."[5]

Einstein had a competent faculty of thinking indeed and he was not going to "fall into it." The problem had become particularly acute after 1905, when Einstein discovered that, in special relativity, no information

[4] Newton was born on 25 December 1642, according to the Julian calender still used in England at that time, but on 4 January 1643, according to the Gregorian calender that had already been adopted in the rest of Europe at that time. It is often said that Galileo died in the same year in which Newton was born, but this is not true when the two events are placed on the same calender. Galileo died on 8 January 1642 (Gregorian calender), which falls in 1641 by English reckoning, since 25 March was taken as the first day of its calender.

[5] I. Newton, in *Four Letters from Sir Isaac Newton to Doctor Bentley Containing Some Arguments in Proof of a Deity*, R. and J. Dodsley, London 1756.

could be transferred at a speed faster than light. The concept of forces acting instantaneously at a distance was incompatible with special relativity.

Starting from these considerations and following a path that turned out to be long (1907–1915) and rugged (because of complex mathematics), Einstein reached a new formulation of the theory of gravity: *general relativity*. According to general relativity, a mass creates a distortion of space and time. Effectively, it "curves" space and time, modifying the fundamental geometric properties to which we are accustomed in "flat" space. In curved space, the motion of a free body is not along straight lines, but it follows the slopes and hills of this distorted space. According to Einstein, geometry replaces the force of gravity, in the sense that the trajectory followed by a free body in curved space exactly coincides with what we perceive as the trajectory of a body subjected to gravity in flat space. Therefore, we interpret gravity as a force, but it is just an effect of the intrinsic properties of space. Rather than being an external force, gravity is the result of a mutual reaction between matter and the geometry of space. Matter modifies space, making it curved; in turn, the space curvature modifies the motion of matter. A single fundamental equation – Einstein's equation – describes the dynamical relation between matter and geometry.

Einstein's general relativity implies a deep revision of fundamental concepts, such as space, time, force and gravitation. In my view, it is the most elegant and captivating scientific theory ever proposed. But general relativity is not just a reformulation of Newton's theory. It predicted new effects – like the anomalous precession of Mercury's perihelion and the bending of light by mass – that have been spectacularly confirmed by observations. General relativity is, at least until the next conceptual revolution occurs, the accepted theory of gravity.

Electromagnetic forces

> Talent is like electricity. We don't understand electricity. We use it.
>
> Maya Angelou[6]

The electric properties of amber rubbed on fur and the magnetic properties of some iron minerals have been known since antiquity. But the first systematic studies of these effects had to wait until the time of William Gilbert (1544–1603), personal physician to Queen Elizabeth I and to King James (VI of Scotland, and I of England). In 1600, Gilbert

---

[6] Maya Angelou, in *Black Women Writers at Work*, ed. C. Tate, Continuum, New York 1983.

introduced the word *electricus* (from the Latin *electrum*, amber), while the word *magneticus* derives from *magnitis lithos*, the mineral extracted in the Greek region of Magnesia, known today as magnetite.

The initial progress in the understanding of electric and magnetic properties was altogether rather slow. During the 18th century, curious machines were built to show the marvels of electricity in elegant salons and in public exhibitions. Later, romantic souls were fascinated by the aura of mystery of electrical phenomena and were captivated by the image of an untamed nature unleashing its power. The poet Percy Shelley, during his university studies at Oxford, read avidly "treatises on magic and witchcraft, as well as those modern ones detailing the miracles of electricity and galvanism". He gathered his friends "discoursing with increasing vehemence of the marvellous powers of electricity, of thunder and lightening, describing an electrical kite that he had made at home, and projecting another and an enormous one, or rather a combination of many kites, that would draw down from the sky an immense volume of electricity, the whole ammunition of a mighty thunderstorm; and this being directed at some point would there produce the most stupendous results."[7]

But the stupendous results of electricity could easily turn into tragedy. While attending a meeting of the St Petersburg Academy of Sciences in 1753, the physicist Georg Wilhelm Richmann (1711–1753) heard thunder announcing an approaching storm. He rushed home to measure the intensity of electricity in lightning bolts, by means of an instrument connected to a metal rod. The lightning came and, without leaving time for the unfortunate Richmann to complete his measurement, instantly electrocuted him.

Eventually science brought some order to this inchoate subject, and the understanding of electrical and magnetic phenomena culminated with the famous Maxwell's equations. James Clerk Maxwell (1831–1879) entered the University of Edinburgh at the age of 16, and three years later moved to the University of Cambridge. The story goes that, on his arrival in Cambridge, he was told that there would be a compulsory religious service at 6 a.m. "Aye," replied Maxwell in his strong Scottish accent, "I suppose I could stay up that late." In that respect, he was like many physicists today. After all, a well-known textbook in physics starts with the definition: "Physics is what physicists do late at night."[8] However, in many other respects, Maxwell was a unique physicist.

Maxwell tackled the problem of electrical and magnetic phenomena starting with an analogy between the laws of fluid dynamics, which describe the flow of liquids, and those of electricity and magnetism, which

[7] T.J. Hogg, *The Life of Shelley*, Moxton, London 1858.
[8] J. Orear, *Fundamental Physics*, Wiley, New York 1967.

describe the flow of charges in electric currents. He then decided to use, as main variables for his equations, the electric and magnetic fields.

The concept of *field*, originally introduced by Michael Faraday, is fundamental to modern physics. A field associates with each point in space a physical quantity, which can be expressed as a number or as a set of numbers. A hiking map showing altitude contour lines can be viewed as an example of a field. A value of the elevation is associated with each point in the map. A red-hot piece of metal can be interpreted as a field too. The colour hues varying along its surface show the value of the temperature, point by point.

The electric and magnetic fields describe, at each point in space, the electric and magnetic forces exerted on a hypothetical particle placed at that point. It may seem that, with this definition, the field is just a convoluted and superfluous concept that only replaces the more intuitive notion of force. But there is more to it than that. The introduction of the field separates two different physical effects: what produces a force from what the force acts upon. In other words, the electric and magnetic fields summarize the information about all the charges and currents present in the system, about their position and their variation with time, but it does not depend on the body on which the force acts.

Maxwell, relying heavily on results from his predecessors, obtained four equations that determine the values of the electric and magnetic fields at each point in space and how they change with time, for any given configuration of electric charges and currents. The equations showed a deep parallelism between electricity and magnetism, such that the two concepts could be merged into a single entity. For this reason today we always refer simply to the *electromagnetic force*.

"From a long view of the history of mankind – seen from, say, ten thousand years from now – there can be little doubt that the most significant event of the 19th century will be judged as Maxwell's discovery of the laws of electrodynamics,"[9] says Richard Feynman. The lightning unleashed by a storm, the alignment of a compass needle, the electric current flowing in a computer microchip are different phenomena explained by the same physical laws describing the electromagnetic force.

Maxwell's discovery revealed an unexpected and truly remarkable result. His equations predicted the existence of waves generated by the mutual oscillations of electric and magnetic fields. Maxwell was able to compute the propagation speed of these waves, and the result was really surprising. The speed obtained by him coincided, within experimental accuracy, with the speed of light, which was known both from laboratory measurements and from astronomical techniques. In 1865, he published

[9] R.P. Feynman, *The Feynman Lectures on Physics*, Addison-Wesley, Reading 1964.

an essay in which he wrote: "This velocity is so nearly that of light, that it seems we have strong reasons to conclude that light itself (including radiant heat, and other radiations if any) is an electromagnetic disturbance in the form of waves propagated through the electromagnetic field according to electromagnetic laws."[10]

Thus the synthesis of electricity and magnetism led to the understanding of the nature of light. According to Maxwell's theory, light is an electromagnetic wave. Its nature is identical to the waves picked up by radios, or those produced in microwave ovens, or the X-rays used for medical diagnosis. These different phenomena have the same physical origin, and the laws describing the electromagnetic force can properly explain all of them.

Maxwell's equations provided the key element to settling an old controversy that had lasted since the beginning of optics: is light made of particles or of waves? Newton opted for a corpuscular interpretation, because obstacles stop light rays, while waves could get around them. Huygens voted for waves, because he observed interference patterns in light. Seven years after Maxwell's death, Heinrich Hertz (1857–1894) produced in the laboratory electromagnetic waves, giving full confirmation of Maxwell's interpretation. At that moment any reasonable doubt was dispelled: light is a wave.

There is a paradoxical aspect in Hertz's experiments, proving once again how unpredictable and haphazard is the way in which science advances. In proving that light is a wave, Hertz planted the seed for the proposition that light is made of particles. In fact, during his experiments on electromagnetic waves, Hertz discovered the phenomenon known as the *photoelectric effect*. This phenomenon presented several puzzling features incompatible with the classical theory of electromagnetism. Then Einstein came up with a satisfactory explanation of the experimental result. Einstein's explanation, however, required a surprising assumption: light is made of particles.

The result was really shocking, because it contradicted all data that gave credibility to the wave interpretation of light. Einstein himself was not able to give any justification to resolve this contradiction and he admitted: "I insist on the provisional character of this concept which does not seem reconcilable with the experimentally verified consequences of the wave theory."[11] But Einstein was right, and later experiments have confirmed that light is indeed made of particles, which today are called *photons*. This is a fundamental result and Einstein was

---

[10] J.C. Maxwell, *A Dynamical Theory of the Electromagnetic Field*, Philosophical Transactions of the Royal Society of London 155, 459–512 (1865).

[11] A. Einstein, in *First Solvay Congress*, ed. P. Langevin and M. de Broglie, Gauthier-Villars, Paris 1912.

rewarded with the Nobel Prize not for relativity – his best-known theory – but for his explanation of the photoelectric effect.

In the confusing results of the photoelectric effect, quantum mechanics was showing its true nature as a theory that defies common intuition. Physicists had to resign themselves to the idea that, in quantum mechanics, the familiar concepts of particle and wave are intrinsically ambiguous and not mutually exclusive. Just as good and evil simultaneously reside in Dr Jekyll and Mr Hyde, showing sometimes one face, sometimes the other, so light can be both particle and wave at the same time. Both Newton and Huygens were right in a sense, and both of them were wrong.

The result that electromagnetic waves are made of photons led to a new interpretation of the electromagnetic field and, hence, of the electromagnetic force. An analogy can help to illustrate the situation. A multinational company has branch offices spread around the world. However, no office operates independently, but only under the influence of all the others, exchanging information about the policy to adopt. A computer network system has been set up to facilitate communication between offices. This network system is a global entity covering the whole world. However, the actual communication occurs through single email messages sent from one office to another. Many email messages are continuously exchanged through the network system. So the global activity of the company is ultimately determined by individual email messages, though in very large numbers.

In this metaphor the branch offices are the various electric charges distributed in space, and the computer network is the electromagnetic field, superintending the behaviour of the system and communicating the force. The email messages are like photons: they are bits of exchanged information. Although the electromagnetic field is a global entity covering all space, it is actually made of particles – the photons. Therefore the electromagnetic force is caused by a continuous exchange of photons between electric charges. This is a really sensational result: not only is matter made of particles, but force is the result of particles too.

Weak force

> In a just cause the weak will overcome the strong.
>
> Sophocles[12]

The story of the discovery of the weak force starts on a cloudy Parisian day in February 1896. Henri Becquerel (1852–1908, Nobel Prize 1903) was engaged in experiments on phosphorescence. He wrapped

---

[12] Sophocles, *Œdipus Coloneus*.

a photographic plate inside a thick layer of black paper, and arranged on top of it a metallic object – a medal shaped like a Maltese cross. Finally, he placed on top of everything a sheet of potassium uranyl sulfate, a uranium salt prepared by him. The first stage of the experiment consisted of exposing his device to sunlight. Then he would develop the plate to observe the photographic image produced by the radiation coming from the phosphorescent material. This radiation, like X-rays, could penetrate the paper, but not the metallic object. The final image would be a photograph, in negative, of the cross.

The sky in Paris remained overcast for several days and, because of lack of sunlight, Becquerel was forced to postpone his experiment. In the meantime, he kept his device locked inside a dark cupboard. Either struck by a prophetic intuition or simply tired of waiting, he decided to develop the photographic plate, although the phosphorescent material had never been exposed to light. To his amazement, he saw an image of the Maltese cross printed on the photographic plate.

This meant that the uranium salt, although never exposed to a light source, was emitting an invisible radiation able to penetrate through the

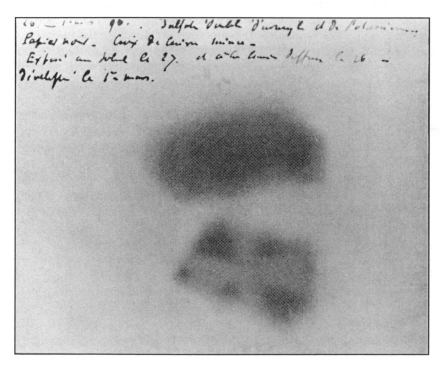

Figure 3.1 The photographic image of the Maltese cross with Becquerel's handwritten notes.

Source: AIP Emilio Segrè Visual Archives / William G. Myers Collection.

black paper. In Becquerel's interpretation, the effect was caused by an "invisible phosphorescent radiation emitted with a persistence infinitely greater than the persistence of luminous radiation."[13] Becquerel had discovered *radioactivity*.

Curiously, in the same month of the same year, Silvanus Thompson (1851–1916) obtained identical results, favoured by English winter weather and the smog layer of Victorian London. Just as for Becquerel, Thompson's apparatus showed an image although it had never been exposed to sunlight. However, Thompson delayed publication of his findings and he misinterpreted the results, attributing the effect to phosphorescence. He did not recognize the evidence for a new form of radiation, thereby missing his appointment with history.

Later studies showed that there are three kinds of radiation emitted by atomic nuclei, which we call *alpha*, *beta*, and *gamma radioactivity*. Alpha radiation is the emission of nuclear fragments from unstable nuclei. These fragments are the alpha particles, which we have already encountered in Rutherford's experiments, formed by two protons and two neutrons bound together. Following an alpha emission, nuclei have to restore equilibrium between their mass and electric charge, and undergo a process called beta radiation. In beta radioactivity, a neutron is transformed into a proton, with the emission of an electron. Finally, gamma radiation is an electromagnetic radiation that is emitted by the nucleus to free some of its energy, usually following an alpha or beta process. Our story here refers only to beta radiation, which is indeed the agent responsible for the image of the Maltese cross observed by Becquerel.

Beta radioactivity presented a puzzle. In all other forms of radioactivity, the energy carried by the radiation was equal to the energy lost by the nucleus. This is fairly intuitive. When you go shopping, the money that you have spent (energy carried out by radiation) must be equal to the money that has disappeared from your wallet (energy lost by the nucleus). It could not be otherwise.

Nonetheless, for beta radioactivity it was indeed otherwise. Many experiments, culminating with the measurements made by Charles Ellis (1895–1980) and William Wooster (1903–1984) at the Cavendish Laboratory in 1927, showed that beta radiation emitted from nuclei was carrying variable amounts of energy. Beta radiation is a stream of electrons, and each electron was measured to have a different energy. It was like firing many shots with the same gun and finding that each bullet has a very different speed. Some of the bullets emerge slowly out of the barrel, some zoom like rockets. How is this possible?

The question reached the desks of theorists, generating a controversy, especially between Niels Bohr and Wolfgang Pauli (1900–1958, Nobel Prize 1945). Bohr reacted to the situation with a revolutionary proposal: at the atomic level, energy is conserved only on average, but not in each

---

[13] H. Becquerel, *Comptes Rendus* 122, 501 (1896).

Figure 3.2 Niels Bohr (left) and Wolfgang Pauli at the Solvay Conference in 1948.
Source: Pauli Archive / CERN.

single process. Therefore, in beta radioactivity, electrons pop out of the nucleus with random energies. In a period in which relativity and quantum mechanics were overthrowing all known principles, nothing in physics seemed sacred anymore. Even the beloved principle of energy conservation – a cornerstone of classical physics – could be demolished, according to Bohr. But Pauli didn't agree: "With his considerations about a violation of the energy law Bohr is on a completely wrong track."[14] Bohr continued to pursue his idea and even imagined that energy violation in beta radioactivity could explain the apparently eternal energy production in stars. But Pauli retorted: "Let the stars radiate in peace!"[15]

Pauli was known for his sharp wit, for his mocking sarcasm and for his loud laugh. But he was also very highly considered and admired by his colleagues for his unique talent and encyclopaedic knowledge. He published his first scientific article, on general relativity, when he was eighteen, a few months after graduating from high school in Vienna. He spent most of his scientific career in Zurich, though travelling often, going wherever he could to discuss physics. When he was speaking in public, he walked incessantly to and fro in front of the blackboard.

[14] W. Pauli, letter to O. Klein, 18 February 1929, in *Wolfgang Pauli, Scientific Correspondence*, ed. A. Hermann, K. von Meyenn, V. Weisskopf, Springer, New York 1979.
[15] W. Pauli, letter to N. Bohr, 17 July 1929, in *Wolfgang Pauli, Scientific Correspondence*, ed. A. Hermann, K. von Meyenn, V. Weisskopf, Springer, New York 1979.

During physics discussions, he used to oscillate his body, as if reciting Hasidic prayers.

Theoretical physicists are often ribbed for having little practical sense or ability to handle laboratory instruments. Pauli excelled as a theoretical physicist also in this respect. He was even credited with the so-called "Pauli effect": the mysterious phenomenon according to which, as soon as he crossed the threshold of a laboratory door, instruments broke into pieces or stopped working for unknown reasons. One day, while James Franck was engaged in an experiment in his laboratory in Göttingen, a sophisticated piece of equipment exploded in an inexplicable way. Franck wrote to Pauli describing the incident and confessing that he had never understood the cause. If only Pauli had been present, he wrote, he could have blamed it on Pauli's metaphysical effect. Pauli checked his diary and was very amused to find out that, at that very moment, he had indeed been at the Göttingen station, waiting for a train connection during a trip between Zurich and Copenhagen.

One day a group of physicists decided to play a practical joke on Pauli. During a conference, they connected the chandelier of the meeting room to a device, concocted in such a way that Pauli, upon entering the room, would involuntarily start the mechanism, making the chandelier fall. At that point, everyone would shout in amazement, invoking the mysterious Pauli effect as the only possible explanation of the incredible accident. Pauli entered into the room, unaware of the conspiracy, but the chandelier didn't fall, because the device didn't work. The Pauli effect had struck once again.

Pauli addressed the problem of beta radioactivity following a different strategy from Bohr. He made the hypothesis that the total energy emitted in beta radiation is always the same. However, only a part of the energy is carried by the electron, while the rest is carried by a new particle – the *neutrino* – with zero electric charge and zero mass (or at least a mass very much smaller than the proton mass). This particle is completely invisible to experimental detectors, because it is nearly insensitive to electromagnetic force. Thus, according to Pauli, experiments measure only part of the actual energy produced by beta radioactivity. This explains the strange experimental observations on the electron energy in beta radiation.

We can rephrase Pauli's explanation with the help of the analogy used previously. All shots fired by a gun have the same energy. Suppose however that the gun does not shoot just one single bullet each time, but two bullets together. One of the two bullets is "visible" (the electron) because it leaves a mark in the target, while the other is "invisible" (the neutrino) because it goes through the target without leaving any trace. The energy of the shot is shared between the two bullets in a random way. Since we are able to see only one of the two bullets, we are deceived and led to believe that some of the energy has disappeared.

Figure 3.3 Wolfgang Pauli during a lecture in Copenhagen in 1929.
Source: Pauli Archive / CERN.

Today physicists are more accustomed to inventing new particles, but in those days Pauli's hypothesis seemed very radical. Actually Pauli did not have the courage to publish his idea in a scientific article, but only described it in a letter sent to a physics meeting on radioactivity at Tübingen, in which he could not participate because had been invited to a ball by the Italian students in a Zurich hotel. Pauli addressed the letter "Dear radioactive ladies and gentlemen," and he explained that his hypothesis was a "desperate way out"[16] of the problem of beta radiation.

Even to Pauli himself the idea seemed like a wild guess. In October 1931, he participated in a conference in Rome. He later said that he held two awful memories of it. The first was that he had to shake hands with

[16] W. Pauli, letter to the Tübingen meeting, 4 December 1930, in *Wolfgang Pauli, Collected Scientific Papers*, ed. R. Kronig, V. Weisskopf, Interscience, New York 1964.

Mussolini. The second was that he had to surrender to the insistence of his colleagues and explain his neutrino hypothesis.

Later in life he described his hypothesis as "that foolish child of the crisis of my life – which also further behaved foolishly."[17] Those were not easy years for Pauli: after the suicide of his mother and his divorce from his first wife, the German dancer Käthe Deppner, Pauli fell into a state of depression. He consulted the psychoanalyst Carl Jung and underwent therapy for almost four years. Jung documented the analysis of hundreds of Pauli's dreams in his book *Psychology and Alchemy*.

In his letter addressed to the Tübingen meeting, Pauli referred to the new particle with the name "neutron". This is because Chadwick's neutron had not, at that stage, been discovered. The name "neutrino" was coined jokingly by Enrico Fermi (1901–1954, Nobel Prize 1938) when, during a seminar in Rome, he was asked if the two particles were the same. "No," replied Fermi, "Chadwick's neutrons are large and heavy. Pauli's neutrons are small and light; they must be called neutrinos."[18] Of course the pun is lost in the English translation: in Italian "neutrino" is the diminutive of "neutron" – "little neutron".

However, the confusion was not just a matter of names, but also of the roles played by these particles. If electrons and neutrinos are emitted from nuclei in beta radioactivity, are they nuclear constituents as much as protons and neutrons? Where are the electrons and neutrinos before being emitted? Pauli originally thought that neutrinos behaved like little magnets and that they are trapped inside the nucleus by electromagnetic forces, but the idea was incorrect.

Eventually, Fermi gave the definitive explanation. He found inspiration in the emerging new ideas of quantum field theory, which had been proposed to understand photon emission and absorption by nuclei. As we will see in more detail in Chapter 4, in a quantum field theory, the concept of force is replaced by interactions among particles. Particles are created or destroyed at various points in space, as a consequence of their interactions. The force is the result of particle exchange and therefore, ultimately, of particle interactions. For instance, because of electromagnetic interaction, photons can be emitted or absorbed by atoms. Nevertheless, photons are not "constituents" of the atom.

Fermi argued that beta radiation could be explained in a way very similar to electromagnetic radiation. He introduced the hypothesis of the existence of a new form of interaction involving neutrons, protons, electrons, and neutrinos. Because of this interaction, a neutron is transformed into a proton with the emission of an electron and a neutrino. In this way, Fermi invented a new force, which was later called the *weak force*, responsible for beta radioactivity.

---

[17] W. Pauli, letter to M. Delbrück, 6 October 1958, in *Wolfgang Pauli, Scientific Correspondence*, ed. A. Hermann, K. von Meyenn, V. Weisskopf, Springer, New York 1979.

[18] G. Gamow, *Thirty Years that Shook Physics*, Anchor Books, New York 1966.

Figure 3.4 *"I ragazzi di Via Panisperna"*: the group of young physicists of the University of Rome led by Fermi. From left to right: Oscar D'Agostino, Emilio Segrè, Edoardo Amaldi, Franco Rasetti, Enrico Fermi.

Source: Archivio Amaldi/Dipartimento di Fisica/Sapienza, Università di Roma.

Fermi was very pleased with his theory and he immediately invited his friends and colleagues Emilio Segrè, Edoardo Amaldi, and Franco Rasetti to come to his apartment. Sitting on his bed, he read to them the article he had written, receiving their enthusiastic comments. Fermi confidently sent his manuscript to the scientific journal *Nature* for publication. But the article was rejected with the justification that "it contains speculations too remote from reality to be of interest to the reader."[19] Eventually the article was published in another journal, and Fermi's theory quickly received full recognition. Actually, the article was published in two journals (one in Italian and one in German), because Fascist rules required that Italian scientists publish in Italian. Since an article in Italian was sure to go completely unnoticed by the international scientific community, Fermi made a double publication.

In spite of the connection noticed by Fermi, there are important differences between the electromagnetic and the weak force. The electromagnetic force does not change the nature of the particle upon which

[19] F. Rasetti, in *E. Fermi, Note e Memorie (Collected Papers) vol. I, Italia 1921–1938,* Accademia Nazionale dei Lincei and The University of Chicago Press, Rome and Chicago 1962.

it acts. In other words, an electron exerts an electromagnetic force on other charges by exchanging photons, but the photon emission does not modify the identity of the electron. Instead, the weak force has the property of transforming particles. In the process of exerting a weak force, the particle itself gets transformed. In particular, in beta radioactivity, a neutron that emits an electron–neutrino pair changes its identity and becomes a different particle – a proton.

The second difference lies in the range at which the force can act. An electric charge exerts a force even at very large distances, and that is why electromagnetism can produce macroscopic phenomena in our world. Instead, the weak force acts only at very small distances. To feel the effects of the weak force, two particles must be at a distance of less than about 10^{-18} metres. This is obviously a distance much too small for us to have a direct sensory perception of the weak force.

Nevertheless, the weak force isn't just a parlour game for particles, it does give effects visible in our world: as visible as the sun, in fact. In a sense, Bohr was right to believe that beta radioactivity was at the origin of the shining of stars, although the explanation had nothing to do with energy violation. Thermonuclear processes driven by the weak force are what make our sun, as well as all other stars, shine.

The third difference between weak and electromagnetic forces rests in their intensities. The neutrino, in contrast with what Pauli first conjectured, interacts with matter only through weak forces. For this reason the neutrino is incredibly elusive and hard to detect by our instruments. For a neutrino, the whole earth is an almost perfectly transparent medium. Actually, only a block of iron as thick as the distance between the earth and Alpha Centauri can be sure to stop a neutrino emitted by beta radioactivity. It is for good reason that this force is called weak.

Nevertheless neutrinos can be experimentally observed. To compensate for the extremely small chance of detecting an individual neutrino, experimentalists use very intense neutrino fluxes, since the total detection probability grows with the number of incoming particles. In 1956 Clyde Cowan (1919–1974) and Frederick Reines (1918–1998, Nobel Prize 1995) obtained the first experimental evidence for the existence of neutrinos at the nuclear reactor of Savannah River, in South Carolina. Soon after their discovery, the two American scientists rushed to send a telegram to Pauli. Pauli consumed a case of champagne with his friends in celebration and then replied with a short letter: "Thanks for message. Everything comes to him who knows how to wait. Pauli."[20]

---

[20] F. Reines, *The Detection of Pauli's Neutrino, in History of Original Ideas and Basic Discoveries in Particle Physics,* ed. H.B. Newman and T. Ypsilantis, Plenum Press, New York 1996.

Strong force

> Only strong personalities can endure history, the weak ones are extinguished by it.
>
> Friedrich Nietzsche[21]

Once it was established that the nucleus is made of protons and neutrons, but does not contain any electrons, the problem was to understand what binds these particles together inside atomic nuclei. The force of gravity was out of the question: it is much too weak for nuclear particles. The electromagnetic force works in the wrong way: it repels protons, contributing to the disintegration of the nucleus, certainly not to its cohesion. After Fermi's discovery, Werner Heisenberg and others tried to use the weak force to explain nuclear stability, but every attempt failed.

Then it became necessary to assume the existence of a new kind of force. This hypothetical new interaction was known as the *strong nuclear force*, because its intensity inside the nucleus had to exceed all other known forces. Not much was known about the strong force, other than a peculiar property. Measurements of proton and neutron diffusion had shown that the strong force acted roughly in the same way on both kinds of particles, and this property was called *charge independence*. But the origin of the strong force remained a total mystery.

In 1934, the Japanese theoretical physicist Hideki Yukawa (1907–1981, Nobel Prize 1949) had the idea that ensured him a place in the history of science: "The nuclear force is effective at extremely small distances. My new insight was that this distance and the mass of the new particle are inversely related to each other."[22] To better understand the meaning of Yukawa's words, let us consider a simple analogy.

One winter day two squabbling brothers go outside to play. A quarrel starts and the two boys begin to throw snowballs at each other. The snowballs exert a repulsive force between the children that pushes them apart. Since the two brothers can throw snowballs a long way, the force persists even at large distances. Calm is restored, but not for long. New accusations fly, these turn to insults, and finally the game becomes more violent. Leaving snowballs aside, the two brothers hurl against each other some sandbags that were found nearby. The sandbags, much heavier than snowballs, cannot be thrown very far. The repulsive force exerted by the sandbags hitting the mischievous boys is very effective, but it acts only at relatively small distances.

The heavier the exchanged object, the shorter is the range of the force. The same thing happens for particles. The electromagnetic

---

[21] F.W. Nietzsche, *Thoughts out of Season, Part 2* (1874).

[22] H. Yukawa, *"Tabibito" (The Traveler)*, World Scientific, Singapore 1982.

force – argued Yukawa – can act at any distance and has an infinite range, simply because the photon has no mass. But if the photon, which is the mediator of the electromagnetic force, had a mass, then the force would act only at short distances. Thus – concluded Yukawa – the particle that mediates the strong force must be heavy, and this explains why the force between protons and neutrons acts only at nuclear distances.

Snowballs flying through the air mediate a repulsive force between the two boys. Photons exchanged by electric charges mediate electromagnetic forces that can be repulsive (for same-sign charges) or attractive (for opposite-sign charges). In much the same way, Yukawa's particles exchanged between protons and neutrons exert an attractive force that keeps the nucleus together. However, while photons have no mass, Yukawa's particles are heavy and so, as in the case of the heavy sandbags, their effect extends only to relatively small distances. The strong force is very intense, but can be felt only up to nuclear distances and not beyond.

Using experimental data on nuclear interactions, Yukawa was then able to compute the mass of the hypothetical particle responsible for the strong force. According to his calculation, the mass of the new particle had to be about 200 times the electron mass. This particle was later called the *meson* (from the Greek *mésos*, in the middle), because its mass is intermediate between that of the electron and the proton. Now the issue was to see whether Yukawa's meson actually existed.

The most powerful particle accelerator in the universe is the universe itself. Cosmic rays, mostly consisting of protons and nuclei, are accelerated by various astronomical sources up to enormous energies, and continuously bombard our atmosphere, penetrating down to the earth's surface. Cosmic rays were the first resource used by physicists to search for new particles whose production requires high-energy collisions.

To detect particles produced by cosmic rays hitting the atmosphere, physicists used *cloud chambers*. Charles Wilson (1869–1959) invented this instrument in Cambridge, while studying cloud formation in meteorology. Cloud chambers are containers filled with gas under special critical conditions, such that a charged particle passing through it triggers the condensation of the gas into miniscule droplets. These droplets form a track that makes the particle trajectory visible, in much the same way that jet aeroplanes leave contrails in the sky. The mass of the particle can be inferred from the thickness of the track. The charge of the particle is measured by the bending of the track under the effect of a magnetic field.

In 1936, Carl Anderson, the physicist who discovered the positron, and his student Seth Neddermeyer (1907–1988) were working at the California Institute of Technology, analysing cosmic rays with cloud chambers. Among the tracks, they discovered a new particle. It had the same charge as the electron, and its mass was measured to be in the range between 100 and 400 times the electron mass, with 200 being the most probable value. A sensational discovery: the meson hypothesized by Yukawa really exists.

"Subtle is the Lord, but malicious He is not."[23] Einstein uttered these famous words in 1921, during his first visit to Princeton, as a comment on rumours of a new experimental measurement that appeared to falsify special relativity. These words were later engraved above the fireplace of Fine Hall, the old Mathematics Department of Princeton University. Less well known is that Einstein, in one of those moments of frustration not uncommon in the activity of theoretical physics, confessed to a colleague: "I have second thoughts. Maybe He is malicious."[24] As we will soon see, the story about the meson discovery gives credence to the latter opinion.

There were early suspicions that something was wrong with the meson discovery. But then the war came, and people were preoccupied with other more pressing issues. The definitive evidence of a problem in Anderson's results came from experiments completed in 1946 by Marcello Conversi (1917–1988), Ettore Pancini (1915–1981) and Oreste Piccioni (1915–2002). These experiments started during the war in the basement of a high school in Rome. The place was chosen because it was sufficiently close to the Vatican to reduce the risk of being bombed. (In modern terminology, this is called background noise reduction.) Electronic components for the experimental apparatus were bought on the black market during Nazi occupation of Rome. Pancini joined the other two physicists only at the end of the war, because he had been previously engaged on commanding resistance brigades in northern Italy. The experiments of the Italian group showed that the particle discovered by Anderson interacted only very weakly with the atomic nucleus and so it could not be Yukawa's meson, which is the messenger of the strong nuclear force.

In 1947, during the Shelter Island conference – one of the most eventful and famous conferences in the history of physics – Robert Marshak (1916–1992) (later in a joint publication with Hans Bethe) proposed a possible explanation of the dilemma. There are two mesons, claimed Marshak: the *muon*, discovered by Anderson, and the *pion*, which is just a new name for the particle proposed by Yukawa. The muon and the pion have nothing to do with each other and they just happen to have roughly the same mass for no good reason other than a malicious coincidence chosen by nature to confuse physicists. Both particles can be produced by cosmic rays, but the pion disintegrates so quickly that it can be only observed at high altitudes. On the other hand, the muon can reach the earth's surface.

[23] A. Pais, *Subtle is the Lord*, Oxford University Press, Oxford 1982.
[24] J. Sayen, *Einstein in America*, Crown Publishers, New York 1985.

Marshak did not know that some Japanese physicists had already proposed the same idea in 1942, but scientific communication between Japan and the United States at that time was rather strained, for obvious reasons. Moreover, none of the scientists present at Shelter Island was aware of a very interesting result obtained on the Old Continent.

New photographic methods for detecting particles had been success-fully developed, especially at the University of Bristol. The technique is based on photographic plates (called *nuclear emulsions*), sensitive to high-energy charged particles. Dark tracks corresponding to the particle trajectories are visible after the plate is developed, showing a real photo-graph of particles. The method is so simple that "even a theoretician might be able to do it,"[25] said Walter Heitler, a theoretical physicist who contributed to the development of this technique.

Soon after the war, Giuseppe Occhialini (1907–1993) and Cecil Powell (1903–1969, Nobel Prize 1950) of Bristol University exposed some nuclear emulsions on the Pic du Midi in the French Pyrenees at an altitude of 2877 metres. In this place today there is an observatory, but at that time access was very difficult. Nevertheless, this was no hindrance to the experiment because of a fortunate circumstance. In 1937, after having discovered the delights of the Fascist regime, Occhialini had left Italy and moved to Brazil. However, when Brazil entered the war, he became an enemy alien and took refuge in the Itatiaya Mountains, working there as a rock-climbing guide. Occhialini's expertise, not only in physics but also in rope work, turned out to be very useful in setting up the experiment at the Pic du Midi.

In their experiment, Occhialini and Powell discovered the pion, a particle that is visible only at high altitudes, because it quickly decays into muons. César Lattes (1924–2005), a Brazilian student who followed Occhialini to Bristol, brought the plates to a meteorological station on Mount Chacaltaya, in Bolivia, at an altitude of 5600 metres. This gave the final confirmation to the discovery of the pion. The Bristol group published their results in October 1947, a few months after the Shelter Island conference. Yukawa's intuition was vindicated and it was demon-strated that his meson – the pion – really exists in nature.

But the intricate story of the strong force, full of great discoveries and wrong interpretations, wasn't over. What does the muon have to do with nature? Some particles (protons, neutrons, and electrons) are there to make matter. Others (photons and pions) are there to make forces. Instead, the muon seemed to have no purpose in the scheme of nature. "Who ordered that?" asked Isidor Isaac Rabi, after the identification of the muon.

This was really a good question to ask. After all, Rabi knew well how to go about asking questions. And all the merit is his mother's. "My

---

[25] O. Lock, *The Discovery of the Pion,* CERN Courier, June 1997.

mother made me a scientist without ever intending to," recalls Rabi. "Every other Jewish mother in Brooklyn would ask their child after school: 'So? Did you learn anything today?' But not my mother. 'Izzy,' she would say, 'did you ask a good question today?' That difference – asking good questions – made me become a scientist."[26]

But instead of finding the answer to Rabi's question, physicists kept on finding new particles. Starting in the 1950s, first in cosmic rays and then in accelerators, many new particles were discovered. These were particles disintegrating just 10^{-24} seconds after being produced, particles that had nothing to do with ordinary matter or forces. The list of "elementary" particles was growing so long that the Greek alphabet was running out of letters to name them after. Once Fermi, noticing that a student was surprised at his hesitation in remembering the name of a particle, simply retorted: "Young man, if I could remember the names of these particles I would have been a botanist."[27]

The situation appeared totally chaotic. It became clear that Yukawa's explanation for the strong force, if not wrong, was at least incomplete. "It is not particles or forces with which nature is sparing, but principles,"[28] said the theoretical physicist Abdus Salam. But, at that moment, nobody had the faintest idea of what the principles were.

[26] Quoted in *Great Minds Start With Questions*, Parents Magazine, September 1993.

[27] The student was Leon Lederman, future Nobel Prize laureate for the discovery of the muon neutrino. The episode is related in L. Lederman and D. Teresi, *The God Particle*, Dell Publishing, New York 1993.

[28] Quoted in S. Weinberg, *Dreams of a Final Theory*, Hutchinson, London 1993.

4
Sublime Marvel

———⊃○◇○⊂———

There is only one step from the sublime to the ridiculous.

Napoléon Bonaparte[1]

The highest reward for a physicist is the discovery that different phenomena have a common explanation that originates from a single principle. The modern theory of particle physics is probably the most successful example of this process of synthesis in science. The theory is an authentic Sublime Marvel that describes all known phenomena in particle physics in terms of a single underlying principle. In this chapter a short overview is given of the events that led to the discovery of this theory.

The construction of the Sublime Marvel required, first of all, the identification of the language able to describe the particle world. This language is *quantum field theory*. Then came the understanding of the electromagnetic phenomena in the domain of particle physics. Finally, the emergence of the Sublime Marvel was the result of three stories intertwined in an inextricable way: the discovery of quarks, the unification of electromagnetism with the weak force, and the understanding of the strong force. The synthesis of these three stories into a single theory was a glorious path that united brilliant theoretical ideas and formidable experiments, plus a healthy dose of false tracks and substantial blunders. The Sublime Marvel was not the creation of an individual mind, as Einstein's general relativity, for example, largely was. Instead it was the cumulative result of the imagination, creativity, and genius of an entire generation of physicists who reached a magnificent synthesis of the laws governing the particle world.

---

[1] N. Bonaparte, remark to the French Ambassador in Poland, Abbé De Pradt, in December 1812 on his way back to Paris after the Russian campaign. Quoted in D. De Pradt, *Histoire de l'Ambassade dans le Grand-Duché de Varsovie en 1812,* Pillet, Paris 1815.

Quantum field theory

> I like relativity and quantum theories because I don't understand them.
>
> David Herbert Lawrence[2]

The process of reconciliation between special relativity and quantum mechanics, started by Dirac, reached its climax with the formulation of *quantum field theory*, which has become the modern language to describe the particle world and provide a new and deeper understanding of the actual meaning of *particle*.

In classical physics the space-filling entities called fields are just a convenient way to describe forces, but their real advantage comes when one considers special relativity. The basic reason is that in special relativity the notion of simultaneity does not have an absolute meaning. This can be understood with the help of a simple example. A man sitting on a train unfolds his newspaper, reads an article, and then folds it back up. From his point of view the unfolding and folding of the newspaper occurred at the same place (the train seat) but at two different instants in time. However, another man standing at the railway station sees the same two events happening at different places, because the train is moving away from him. Moral: two events happening at the same point in space but at different instants in time for one observer are separated by a space interval from the point of view of another observer in relative motion.

Special relativity has revealed a complete parallelism between the concepts of space and time. So we are allowed to switch the words "space" and "time" in the moral above and obtain a new assertion: two events happening at the same instant in time but at different points in space for one observer are separated by a time interval from the point of view of another observer in relative motion. In other words, the same two events can be simultaneous for one observer but separated in time for another observer.

While the original assertion makes perfect sense and is quite intuitive, the second assertion seems almost paradoxical, but it is nonetheless true. Our fallible intuition tends to give an absolute meaning to the flowing of time, and statements that seem evident in the case of space become almost absurd in the case of time. But these statements are true, because in special relativity space and time merge into the same concept, and one of the consequences is that the notion of simultaneity has no absolute validity.

---

[2] D.H. Lawrence, *Pansies: Poems*, Martin Secker, London 1929.

The crisis with the concept of simultaneity made action at a distance an untenable idea. Forces cannot act simultaneously at different points, and there must exist some physical entity that carries the force throughout space. The concept of field eliminates altogether any reference to action at a distance, by mathematically implementing a fundamental notion of physics: *locality*.

Locality means that the behaviour of a system depends only on properties defined in its vicinity (in space and time). Martians cannot modify the outcome of collisions at the LHC without sending some intermediate agent from Mars to the vicinity of the collider. Although locality is not a logical necessity of nature, it is a well-established empirical fact observed in physical laws. And, for science, it is a very useful fact indeed. If locality did not hold, the interpretation of any laboratory experiment – such as the oscillation of a pendulum or the radioactive decay of a nucleus – should take into account the position of the planets or the velocity of distant galaxies, and physics would be an inextricable mess. But, luckily, this is not the case.

When you throw a rock into the middle of a lake, you can see the disturbance created on the surface of the water propagating in the form of circular waves. These waves can reach a distant buoy, making it bob up and down. The effect of the rock on the buoy did not occur through action at a distance, but rather through the mediation of the propagating wave. In the same way, fields carry the information of force throughout space, strictly preserving the property of locality. A system reacts only to the action of the fields in its vicinity.

After the rules of quantum mechanics had been successfully applied to the electron, it appeared logical to extend them to an old acquaintance of physics since the time of Maxwell – the electromagnetic field. The mathematical procedure for doing so, called *field quantization*, had already begun to be used in 1926 by Max Born (1882–1970, Nobel Prize 1954), Werner Heisenberg (1901–1976, Nobel Prize 1932) and Pascual Jordan (1902–1980), leading to several surprises. The quantum field – namely the electromagnetic field after the procedure of quantization – did not turn out to be a continuous medium, like the fields imagined by Maxwell, but it decomposed into a series of individual lumps of energy.

To use a metaphor, we can view the quantum field as a vast sea, covering all space. In various places, the sea surges into waves, billows and swells that propagate along the surface. Examined individually, each wave or billow looks like a separate entity but, in reality, they are all part of the same substance – the sea. In the same way, the quantum field contains lumps of energy that propagate in space. These individual lumps of energy are what we call particles. But in reality particles are just an expression of the underlying substance that fills space – the quantum field. Particles are nothing else but a localization of the field energy, just like billows are localized surges in the water level.

This new interpretation of the concept of particle helps to explain some of the puzzles encountered in quantum mechanics. The duality between particle and wave, so mysterious for the pioneers of quantum mechanics, now emerges more clearly from the procedure of field quantization. The electromagnetic field behaves like a wave obeying Maxwell's equations. However, when viewed under the magnifying lens of quantum mechanics, the electromagnetic field no longer looks like a continuous medium, but rather like a swarm of individual entities, each carrying a fixed amount of energy. These lumps of energy are the photons, which behave like particles. Photons, therefore, are a necessary consequence of quantum mechanics applied to Maxwell's electromagnetism.

The procedure of field quantization is actually valid not only for photons, but also for any kind of particle. Take for instance the electron. Electrons were believed to be individual particles. But in reality they are lumps of energy of a quantum field that fills all space. Each kind of elementary particle (electron, photon, and so on) is associated with a different quantum field. Quantum fields are the LEGO pieces that nature plays with to build her wondrous creations. Quantum fields, and not particles, are the primary reality.

In a human population, each individual is different and unique. Some have brown eyes, some blue; some have blonde hair, some black; some are taller than others, some more intelligent. By contrast, electrons are all exactly identical, like an alien population of clones. The same is true for photons or for any other particle: each of them is an exact copy of the others. This is not surprising, in view of the particle interpretation given by quantum field theory. There exists a single entity that describes all electrons: the quantum field of the electron. The quantum field undergoes internal pulsations, concentrating its energy in certain points of space. We observe these lumps of energy as individual electrons, but actually they are a manifestation of a single physical substance.

In a stormy sea many waves and billows are formed, some of them colossal, others tiny, but all of them are made of the same substance: water. In the same way electrons can have different speeds, some of them are very fast, others are slow, but all of them have the same intrinsic properties (such as mass and electric charge) because all of them are manifestations of the same quantum field. A single field describes all the electrons present in the universe.

From the point of view of quantum field theory, the electromagnetic force is the result of the interaction between the electron and the photon fields. It is as if different liquids filled our metaphoric sea. Each liquid produces its own ripples, propagating on the surface of the sea. When waves made from different liquids come in contact, they have a reciprocal effect: some waves disappear, some waves swell by absorbing the energy of others, new waves are produced. The same happens for particles: photons can be absorbed or emitted by electrons; particles can disappear transforming their energy into other kinds of particles. But the interactions between fields are strictly *local*: particles affect each other only at the same point of space and time.

The particle world is an ever-changing environment, where energy is quickly transformed into mass and vice versa, where new particles constantly appear and disappear, like waves in a stormy sea. The LHC is like a spectacularly violent tempest, where two colossal tsunamis clash in a single point. Out of this powerful storm, new waves will be formed, perhaps even of kinds as yet unknown.

A formidable consequence of quantum field theory has been the conceptual unification of matter and force, which in our everyday experience appear so different. The distinction between what exerts the force (the electron) and what transmits it (the photon) is merely a matter of classification. Both matter and force are the result of quantum fields and of their mutual interactions. This is indeed a crucial step in the direction of synthesis and unification, which is the path towards the discovery of the fundamental principles of nature.

Quantum electrodynamics

> The theory of quantum electrodynamics describes Nature as absurd from the point of view of common sense. And it fully agrees with experiment. So I hope you can accept Nature as She is – absurd.
>
> Richard Feynman[3]

The idea of quantum field theory found its first successful application in what is now called *quantum electrodynamics*. This theory is usually referred to by the acronym QED, which stands for Quantum Electro Dynamics, but is also a pun on QED, the abbreviation of *quod erat demonstrandum*, indicating the end of a mathematical proof. QED is the theory that describes electrons, photons, and their mutual interactions, extending Maxwell's laws of electromagnetism to the world of particles, where special relativity and quantum mechanics are essential ingredients.

Unfortunately the theory, soon after being born, started to cause trouble. Instead of behaving with the joyful bliss of a baby, it immediately showed the fickle irritability of a teenager. In some cases calculations were giving results in perfect agreement with experimental data, but sometimes the results were equal to infinity, that is a number larger

---

[3] R.P. Feynman, *QED: The Strange Theory of Light and Matter*, Princeton University Press, Princeton 1985.

than any number you can imagine. This was completely absurd and against all logic.

The initial attitude of many physicists was to retain the successful results and just ignore the absurd ones, assuming that everything that turned out to be infinity must not exist at all. Of course, this attitude was not taken seriously for too long. "Just because something is infinite," theoretical physicists joked, "does not mean it is zero."[4]

The turning point came at the conference at Shelter Island, which we have already encountered in the story of the pion's discovery. This conference, held during 2–4 June 1947 at the Ram's Head Inn on Shelter Island, New York, marks a transition in the history of physics for three main reasons. The first is that it represents the birth of the modern vision of the particle world based on field theory. Richard Feynman later declared: "There have been many conferences in the world since, but I've never felt any to be as important as this."[5] Robert Oppenheimer and John Wheeler made similar comments. The second reason is that, in spite of the strict security measures present at Shelter Island, physicists could go back and discuss science without the incumbent fear of war and the military yoke of the Manhattan Project. Julian Schwinger said: "It was the first time that people who had all this physics pent up in them for five years could talk to each other without somebody peering over their shoulders and saying, 'Is this cleared?' "[6] The third aspect is geographical. The conference marked the shift of the centre of activity in physics from Europe to the USA. This change happened partly because of the racial laws that had forced many of the European protagonists to emigrate, and partly because Europe, prostrated by the war, did not have the economic resources to adequately fuel fundamental research.

In relation to our story, two important experimental results were presented at the conference. Willis Lamb (1913–2008, Nobel Prize 1955), a young American experimentalist who had started as a theoretician studying under the supervision of Robert Oppenheimer, illustrated the first result. Using the technology of microwave radar developed during the war, he had succeeded in measuring a separation between two spectral lines of hydrogen, which according to Dirac's theory should coincide. This separation between spectral lines was later called the *Lamb shift*.

[4] S. Weinberg, *The Quantum Theory of Fields,* vol. I, Cambridge University Press, Cambridge 1995.

[5] R.P. Feynman, interviewed by C. Weiner in 1966, *Archives for the History of Quantum Physics*, Niels Bohr Library, American Institute of Physics, College Park, Maryland.

[6] S.S. Schweber, *QED and the Men Who Made it: Dyson, Feynman, Schwinger, and Tomonaga*, Princeton University Press, Princeton 1994.

Isidor Isaac Rabi (1898–1988, Nobel Prize 1944) presented the second result. He had measured the intensity of the magnetism associated with the intrinsic rotation of the electron, called its spin, finding a value in excess of that predicted by Dirac's theory by only 0.1 per cent. This excess, called the anomalous magnetic moment of the electron, is usually identified by the symbol $g - 2$.

The result reported by Lamb forced theorists to confront the infinities in an open battle; no coward-like retreat was possible any longer. This is because QED was indeed predicting an effect in the Lamb shift, but the contribution turned out to be infinite. The experiment presented at Shelter Island was proving once and for all that something infinite isn't necessarily zero!

The theoreticians did not get disheartened. Discussions started immediately at Shelter Island, under the leading of Robert Oppenheimer (1904–1967), Hans Kramers (1894–1952), and Victor Weisskopf (1908–2002). The first ideas on how to cope with infinities were quickly proposed. On the train ride back from the conference, Hans Bethe (1906–2005, Nobel Prize 1967) completed the calculation of the Lamb shift.

Figure 4.1 Physicists discussing at the Shelter Island Conference in 1947. From left to right: (standing) Willis Lamb, John Wheeler; (sitting) Abraham Pais, Richard Feynman, Herman Feshbach, Julian Schwinger.

Source: AIP Emilio Segrè Visual Archives.

Soon after, Julian Schwinger (1918–1994, Nobel Prize 1965) computed $g - 2$, obtaining a result in superb agreement with the measurement by Rabi. By the end of the 1940s Richard Feynman (1918–1988, Nobel Prize 1965), Julian Schwinger, and Sin-Itiro Tomonaga (1906–1979, Nobel Prize 1965) completed the proof that any physical process in QED can be computed and the result is finite. The battle against infinities was finally won. How was it done?

Imagine that tomorrow is St Valentine's Day. You and your friend David Beckham go out shopping to buy presents for your respective wives. You enter a store and David chooses for Victoria 30 diamond chokers, 50 emerald bracelets, 60 fur coats plus some other expensive items. He keeps careful track of his expenditures, which total some megabillion zillion euros. You pick up a small bouquet of flowers, whose price isn't marked. In the confusion at the checkout counter, all your purchases are rung up together and the total bill amounts to some mega-billion zillion euros. Must you really pay some megabillion zillion euros for a bouquet? Of course not: all you have to do is take the difference between the total bill and David's share, and you find that you must pay only 19 euros and 99 cents.

Something similar happens in calculations in QED. Most of the results of these calculations are equal to colossal numbers (actually infinity). However, these results do not correspond to measurable physical quantities, as much as the total bill above does not refer to what you must actually pay. Once the result of a physical quantity is appropriately expressed in terms of other physical quantities, colossal numbers are subtracted from each other and the result is a perfectly reasonable small number. This procedure is called, in scientific jargon, *renormalization*. For instance, QED gives huge (actually infinite) corrections to the electron mass, to the electron charge, and to the Lamb shift. However, once the Lamb shift is expressed in terms of the total electron mass and charge, huge numbers are subtracted from huge numbers, leaving a perfectly sensible result.

QED allows theoretical physicists to make astoundingly precise predictions about electromagnetic processes occurring in the particle world. Experiments have made tremendous progress too. For instance, the magnets associated with the motions of the muon and the electron have been measured with a relative precision of 6×10^{-10} and 3×10^{-13}, respectively. The latter accuracy is equivalent to determining the earth's circumference with a margin of error of ten microns. These experimental measurements are in superb agreement with the theoretical calculations based on QED.

In spite of the splendid success of QED, in the post-war period many physicists were regarding with suspicion the procedure of renormalization, which appeared to them as a mathematical trick to hide some deep conceptual problem. The presence of infinities was viewed as an indication that quantum field theories were sick structures that would soon die and leave space for new theories, more pleasing to the refined palates of the mathematical physicists. But from a practical point of view, the real limitation of quantum field theories lay in the difficulty of applying them outside the domain of the electromagnetic force. For the weak force, it was not possible to write down a theory where all the infinities could be eliminated. For the strong force, it was possible to formulate such a theory but it was of no practical use, because nobody knew how to deal mathematically with its equations.

For these reasons, during the 1950s and part of the 1960s, the shares of quantum field theory fell rather low on the stock market of theoretical ideas. Theoretical physicists preferred to concentrate in finding alternative theories. As Schwinger later said: "The preoccupation of the majority of involved physicists was not with analyzing and carefully applying the known relativistic theory of coupled electron and electromagnetic fields, but with changing it."[7] The new theoretical proposals (S-matrix, bootstrap, non-local theories, fundamental-length theory, to name but a few) do not play much of a role in today's description of elementary particles, but were the fads of the time. The disenchantment with quantum field theory turned out to be premature, but many things had to be understood before the powerful arsenal of quantum field theory could be fully exploited.

The discovery of quarks

> Do not infest your mind with beating on the strangeness of this business.
>
> William Shakespeare[8]

While QED completely solved the problem of electromagnetism, the situation with the other forces was growing ever more complicated. The problem was certainly not the lack of experimental discoveries, but rather that nature was presenting physicists with an embarrassment of riches. Starting from the 1950s more and more particles were discovered; by the 1960s there were more than a hundred. Any hope that nature was simple at the microscopic level seemed vain.

To impose some order on the subject, physicists classified particles into two families. *Leptons* (from the Greek *leptós*, thin) are particles that do *not* feel the effect of the strong force, but only of the weak force

---

[7] J. Schwinger, in *The Birth of Particle Physics,* ed. L. Brown and L. Hoddeson, Cambridge University Press, Cambridge 1983.

[8] W. Shakespeare, *The Tempest*, Act V, Scene I.

and, in some cases, the electromagnetic force. Actually only three leptons were known at the time: the electron, the muon, and the neutrino. Another lepton, the *tau* (from the initial of the Greek word *tríton*, third), was not discovered until the mid 1970s. Later research showed that there are in fact three neutrinos, associated with the electron, muon and tau, respectively.

The really complicated and ever-growing family of particles was known under the collective name of *hadrons* (from the Greek *hadrós*, thick). Hadrons are particles affected by the strong force – like protons, neutrons, pions, and many others. Incidentally, the "H" of the "LHC" refers to the fact that protons are hadrons.

Hadrons represented a real puzzle. Some of them were found to have the unexpected property of disintegrating rather slowly, in spite of being readily produced by strong interactions. It was like throwing a coin up in the air, and then having to wait for thousands of years to see the coin come back to your hand. Physicists, lacking a real understanding of this strange phenomenon, could think of nothing better than assigning to these particles a new property which they called *strangeness*.

Actually the introduction of strangeness led to a useful clarification. Like stamps collected in an album, hadrons could be classified according to their strangeness and electric charge. When this was done, the arrangement of various groups of hadrons was found to be forming well-defined geometrical figures. If one vertex of the figure remained empty, like the missing stamp in a collection, experiments looked hard for a new particle with the correct properties of strangeness and charge and, unfailingly, they were discovering it.

In 1961, Murray Gell-Mann (Nobel Prize 1969) and Yuval Ne'eman (1925–2006) – an engineer turned physicist, an officer in the Israeli army and later Minister of Science – independently identified the symmetry properties of the hadron structures. This scheme was called the *Eightfold Way* by Gell-Mann, who had a peculiar ability in inventing original, if not queer, names: "The new system has been referred to as the 'Eightfold Way' because it involves the operation of eight quantum numbers and also because it recalls an aphorism attributed to Buddha: 'Now this, O monks, in noble truth that leads to the cessation of pain, this is the noble Eightfold Way: namely, right views, right intention, right speech, right action, right living, right effort, right mindfulness, right concentration.' "[9]

The Eightfold Way is to hadrons what Mendeleev's period table is to atoms. Classification is often a first step towards a deeper understanding of the internal structure. The atoms' arrangement in the period table

[9] G. F. Chew, M. Gell-Mann and A. H. Rosenfeld, *Scientific American*, February 1964, 74.

was explained by their compositeness in terms of protons, neutrons, and electrons. In the same way, Murray Gell-Mann and George Zweig argued independently that the structure of the Eightfold Way was magically reproduced by the hypothesis that hadrons were composed of new entities – *quarks*.

Actually Zweig, who was working at CERN at the time but later moved first to neurobiology and then to the financial investment industry, gave to the hypothetical components of hadrons the name *aces*. But competing with Gell-Mann in terms of finding bizarre names for particles was a lost cause. Gell-Mann introduced the name "quark" during physics discussions, apparently just as a facetious rhyme of "pork". "I had the sound first, without the spelling, which could have been 'kwork'," explains Gell-Mann. "Then, in one of my occasional perusals of 'Finnegans Wake', by James Joyce, I came across the word 'quark' in the phrase 'Three quarks for Muster Mark'. Since 'quark' (meaning, for one thing, the cry of a gull) was clearly intended to rhyme with 'Mark', as well as 'bark' and other such words, I had to find an excuse to pronounce it as 'kwork'."[10] The excuse was rather untenable, but the name "quark" caught on nevertheless.

The hypothesis that hadrons were made of quarks was working perfectly well, save for one drawback: no experiments had ever observed a quark. And yet, quarks should have been lighter than protons and have a fractional electric charge, because they were constituents of protons. However, in spite of active experimental searches, no particle with fractional electric charge was ever discovered. There was no sign whatsoever of the existence of quarks. This seemed a very good reason for the majority of physicists to doubt the physical reality of quarks. They were considered just a practical mnemonic trick to recall the properties of hadrons.

The general incredulity of the reality of quarks is well illustrated by an episode that occurred in 1963, when Gell-Mann made a phone call from California to Victor Weisskopf, previously his doctoral advisor, to discuss physics with him. "I did talk with Viki Weisskopf, then director general of CERN," recalls Gell-Mann, "in the early fall by telephone between Pasadena and Geneva, but when I started to tell him about quarks he said, 'This is a transatlantic phone call and we shouldn't waste time on things like that.'"[11]

Gell-Mann was an extremely influential scientist at the time. A child prodigy, he is a refined linguist and a man of stunningly broad culture. Among his many interests, he is also an expert ornithologist. I was once told that, during a visit to Israel, he was taken on a jeep tour in

[10] M. Gell-Mann, *The Quark and the Jaguar*, Abacus, London 1994.

[11] M. Gell-Mann, in *The Rise of the Standard Model*, ed. L. Hoddeson, L. Brown, M. Riordan, M. Dresden, Cambridge University Press, Cambridge 1997.

the Negev Desert. Suddenly, Gell-Mann excitedly announced to his companions that he heard the shriek of an extremely rare bird. The jeep started in pursuit, following the instructions of Gell-Mann who, ears cupped, was attentively listening to the sound, recognizable only to his expert hearing. After much wandering about in forlorn areas of the desert, the driver stopped the jeep, because he finally understood what Gell-Mann was hearing. It was the belt of the engine squeaking while the old jeep was bouncing on the rocky desert tracks. This story proves that even Gell-Mann isn't infallible, but nevertheless with quarks he wasn't wrong.

Things changed in 1968. Experiments at Stanford Linear Accelerator Center (SLAC) in the USA directed by Jerome Friedman (Nobel Prize 1990), Henry Kendall (1926–1999, Nobel Prize 1990), and Richard Taylor (Nobel Prize 1990) studied the structure of protons by probing them with electron beams. The logic of the experiment was very similar to the probe of the atomic structure made by Geiger, Marsden, and Rutherford more than 50 years earlier. Measuring the deflection of the impinging electrons, the SLAC experiment could infer how the electric charge was distributed inside the proton.

The theoretical physicists James Bjorken and Richard Feynman interpreted the experimental data, and the result was a real surprise. The evidence was that the electric charge of the proton is not uniformly distributed, but concentrated in point-like particles in its interior. This meant that protons are composite particles made up of quarks. "I was always delighted when something esoteric could be made to look so simple,"[12] declared Feynman.

But the most surprising aspect of the theoretical interpretation was that quarks inside the proton appeared not to feel any strong reciprocal force. If quarks were really the constituents of the proton, one would naturally expect a strong binding force able to keep them captive inside its interior. Instead, quarks behaved like free particles, not mutually interacting. It was as if quarks were locked inside a prison with invisible walls. Quarks were perfectly free to move inside the proton, but they could not cross the proton's boundaries. But then, if quarks do not feel strong binding forces, why was no one able to observe quarks in isolation outside the proton?

The electroweak theory

> Our mistake is not that we take our theories too seriously, but that we do not take them seriously enough.
>
> Steven Weinberg[13]

---

[12] R.P. Feynman, as quoted in M. Riordan, *The Hunting of the Quark*, Simon & Schuster, New York 1987.

[13] S. Weinberg, *The First Three Minutes*, Basic Books, New York 1977.

In the early 1960s, Fermi's theory was still providing the current description of the weak force, although the original theory had been much refined. This process of refinement led to its own series of surprises, for it was found that the weak force presented unexpected features when the roles of matter and antimatter were exchanged, or when the coordinates of space were reflected as in a mirror.

But Fermi's theory could not be the final word on the weak force. As photons transmit the electromagnetic force and pions were believed to carry the strong force, it was logical to suppose that some new particle was the messenger of the weak force. This hypothetical particle was referred to with the name *W*. The construction of the theory of the *W* particle and of the weak force presented serious difficulties, but eventually the problem was solved in the context of *gauge theories*. Gauge theories and the problem of the weak force play a crucial role in the research programme of the LHC. For this reason, these concepts will be treated separately in Chapter 8 after all the necessary elements have been introduced. For the moment, it is sufficient to know that gauge theories are generalizations of QED that describe the action of a force mediated by the exchange of particles. While in QED the electromagnetic force is carried by a single kind of particle (the photon), a general gauge theory contains several different kinds of "photon", that is to say several different kinds of force carrier.

In 1967 Steven Weinberg (Nobel Prize 1979) and Abdus Salam (1926–1996, Nobel Prize 1979) identified the correct theory of the weak force, applying new ideas about gauge theories to a model first proposed by Sheldon Glashow (Nobel Prize 1979). Weinberg was actually studying how these ideas could be applied to describe the strong force. But soon he had the right inspiration: "Then it suddenly occurred to me that this was a perfectly good sort of theory, but I was applying it to the wrong kind of interaction. The right place to apply these ideas was not to the strong interactions, but to the weak and electromagnetic interactions."[14] The situation with quarks and hadrons was too muddled, and Weinberg decided to focus on leptons, the particles that are not affected by the strong force. He obtained a theory that wonderfully describes the interactions between leptons and the *W*, reproducing all known features of the weak force.

The Dutch physicist Martinus Veltman (Nobel Prize 1999) firmly believed in quantum field theory, even when fashion had moved elsewhere. He had developed new techniques to confront infinities in gauge theory and assigned to his student Gerardus 't Hooft (Nobel Prize 1999) the task of studying the renormalization procedure of the new theory of the weak force. In 1971 't Hooft, under the guidance of Veltman, showed that all infinities were properly disappearing. After this result was

[14] S. Weinberg, *The Making of the Standard Model,* The European Physical Journal, C 34, 5 (2004).

announced, the new theory of the weak force reached complete acceptability in the eyes of theorists.

A crucial aspect of the new theory of weak interactions was that the photon and the *W* appeared as two different carriers of the same force. A unified theory described simultaneously the electromagnetic and weak forces. More than 100 years after Maxwell, a new step in the unification of forces was made, so that today we speak of a single *electroweak force*.

How can electromagnetic and weak forces be the same entity when they look so different to us? The crucial point is that the weak force appears so feeble only because of its very limited range, but not because of its intrinsic strength. When physicists were able to descend – first intellectually and then experimentally – into the particle world at very short distances, they had the surprise of discovering that weak and electromagnetic forces behave in the same way and are united in a single concept. The name "weak force" is deceptive, because its intrinsic strength is actually comparable to the strength of electromagnetism. But we will have to wait for Chapter 8 to fully appreciate how unification works.

The hypothesis of electroweak unification not only represented conceptual progress, but it also made a clear prediction that could be confronted by experiment. The consistency of the theory was requiring the existence of a new force carrier, beside the photon and the *W*, which became known as the *Z* particle. This particle, contrary to the electrically charged *W*, was predicted to have no charge. The *Z* particle would then mediate a new kind of weak force. Neutrinos were known to interact with matter by transforming themselves into electrons – a process mediated by the *W* and called the *charged current*. But if the *Z* particle really existed, then neutrinos would interact with matter also without modifying their identity and remaining neutrinos – a process called the *neutral current* interaction.

The discovery of neutral currents became a primary experimental goal because it would have provided tangible evidence for electroweak unification. Unfortunately the identification of neutral currents was a very difficult task because neutrinos are extremely elusive particles. The problem was not just to identify neutrinos, but also to measure their footprints as they bounce off matter.

In 1963 André Lagarrigue (1924–1975), Paul Musset (1933–1985) and André Rousset (1930–2001) elaborated a proposal for a neutrino experiment, using a *bubble chamber* as particle detector. Bubble chambers evolved from cloud chambers, which had been used in the discovery of the positron and the muon. Cloud chambers identified particle trajectories through tracks of condensation droplets, whereas particles traversing a bubble chamber leave a trail of small bubbles inside a vessel filled with superheated liquid, which is in an unstable condition on the verge of boiling. The story goes that Donald Glaser (Nobel Prize 1960), the inventor of the bubble chamber, was staring at the columns of bubbles in his glass of beer when he had the idea of this kind of particle detector.

The bubble chamber planned by the French group was enormous for that time: a 4.8 m long cylinder with a 1.9 m diameter, filled with liquid freon. Although just a child's toy by comparison with the LHC detectors, the instrument was considered so big at the time that the director of the École Polytechnique, when shown a drawing of the project, gave it the name *Gargamelle*, the mother of the giant Gargantua in the 16th century novel by Rabelais. Looking out of the window of my office, I can still admire the bubble chamber of Gargamelle, because it is now exhibited on a lawn inside CERN.

The experimental collaboration of Gargamelle grew larger during the construction phase of the instrument and in 1971, when everything was ready for the first data taking, it included 60 physicists from seven different European laboratories. Not very impressive numbers compared with LHC experiments, but it was certainly the first example of such a large scientific collaboration. In December 1972, physicists involved in the analysis of Gargamelle data identified the first clear image of a "neutral-current event" – the scattering of a neutrino off an atomic electron. The discovery created great excitement.

Figure 4.2 The Gargamelle bubble chamber at CERN in 1970.
Source: CERN / Gargamelle Collaboration.

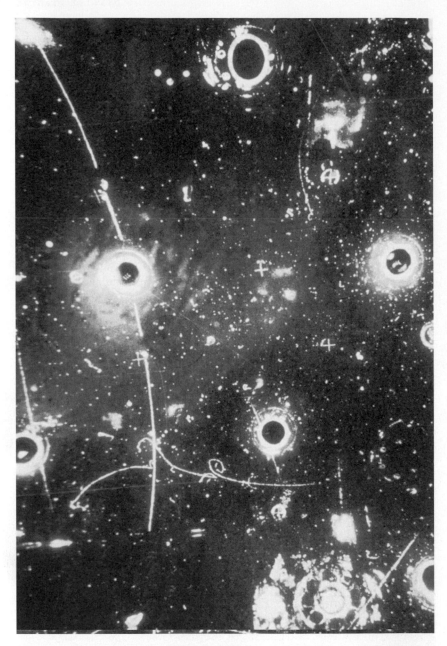

Figure 4.3 The image of a neutral current interaction recorded by Gargamelle in 1973. The track shows an electron that has been kicked off its atomic orbit by an incoming neutrino.

Source: CERN / Gargamelle Collaboration.

Gargamelle was not the only experiment searching for neutral currents. At Fermilab, the particle physics laboratory near Chicago, the collaboration HPWF (Harvard Pennsylvania Wisconsin Fermilab) confirmed the evidence found by Gargamelle. But later, after an upgrade of their instrument, they announced a new result for the ratio between neutral and charged currents, much smaller than before and actually compatible with the total absence of any neutral current. The pressure was on to resolve the conflicting results from the two experiments.

Donald Perkins, of the Gargamelle collaboration, recalls: "The Americans had vastly more experience and know-how.... It is important to understand this legacy of inferiority in considering the attitudes at that time of people in CERN over the Gargamelle experiment. When the unpublished (but widely publicized) negative results from the HPWF experiment started to appear in late 1973, the Gargamelle group came under intense pressure and criticism from the great majority of CERN physicists. Part of this was presumably just prejudice against the technique: people could not believe that such a fundamental discovery could come from such a crude instrument as a heavy liquid bubble chamber.... But equally important, many people believed that, once again, the American experiments must be right. One senior CERN physicist bet heavily against Gargamelle, staking (and eventually losing) most of the contents of his wine cellar!"[15]

Gargamelle physicists stuck to their claims and eventually the discrepancy with the American experiment was resolved. After the upgrade, an imperfect understanding of the instrument's performance led HPWF to misinterpret some neutral-current events as due to charged currents. Once the functioning of the detector was fully understood, HPWF confirmed the findings of Gargamelle. Any remaining doubt had been swept away: neutral currents had been discovered. Some European physicists could not resist the temptation to comment jokingly on the oscillating results from HPWF by claiming that the Americans had identified "alternating neutral currents".

The discovery of neutral currents gave incontrovertible evidence for the unification between electromagnetic and weak forces. Moreover, from the measurement of the ratio between neutral and charged currents, it became immediately possible to obtain a first estimate of the masses of the hypothetical W and Z particles. This opened the experimental hunt for these particles, which, although firmly believed to exist by theorists, had not yet been directly detected.

At that time CERN was planning the construction of a large accelerator colliding electron and positron beams. This accelerator, called LEP (Large Electron–Positron collider), was well suited for the discovery

[15] D. Perkins, in *The Rise of the Standard Model*, ed. L. Hoddeson, L. Brown, M. Riordan, M. Dresden, Cambridge University Press, Cambridge 1997.

of the *W* and *Z* particles, but it would start operating too far in the future for physicists who were very anxious to demonstrate the reality of these particles. Proposals for new proton colliders were made, but the CERN management did not approve these projects for fear that they would delay LEP. In this unsettled situation, Carlo Rubbia (Nobel Prize 1984) came up with a daring and radically different idea.

In 1977 he submitted to both CERN and Fermilab the proposal to convert existing proton machines into accelerators colliding a proton beam against an antiproton beam. Rubbia was convinced about a technique, developed by Simon van der Meer (Nobel Prize 1984), capable of storing and accelerating intense beams of antiprotons. But nobody was certain that this technology was feasible and there was much scepticism about converting brand new machines ready to operate, such as the CERN SPS, into a collider with little chance of it working. "Much of the merit of Carlo Rubbia is to have pushed his ideas with such an untiring determination and in such an adverse context. Not only with determination but also with a clear vision of what they turned out to lead to and with a deep understanding of the machine physics issues at stake,"[16] recalled Pierre Darriulat, a physicist who played an important role in the discovery of the *W* and *Z*.

The basic idea was to shoot a proton beam against a target in order to create antiprotons. For every million incident protons only a single antiproton would be produced. Antiprotons would then be cooled and accumulated in a ring, at a rate of one hundred billion a day. Finally, the antiproton beam would be brought to a head-on collision against a proton beam. Rubbia made a prototype experiment, the Initial Cooling Experiment (ICE), which successfully demonstrated the technology. In 1978 John Adams and Léon Van Hove, the two CERN director generals, took the bold step of approving the proton–antiproton project. "It is very difficult to rewrite history, all events are so intricately linked to each other, but I strongly believe that, if it had not been for Carlo, there would have been no proton–antiproton collider physics in the world for a long time, maybe ever,"[17] stated Darriulat.

Two large detectors, UA1 and UA2, were designed and constructed to observe the particles bursting out of the collisions between protons and antiprotons. After a strenuous tour de force, on 9 July 1981 the first particle collision occurred, only three years after the project approval. A few hours later, particle collisions had already been recorded and publicly shown. Very few people would have believed that all this was possible, let alone in so little time.

---

[16] P. Darriulat, *The Discovery of the W & Z, a Personal Recollection*, The European Physical Journal, C 34, 33 (2004).

[17] P. Darriulat, *ibid.*

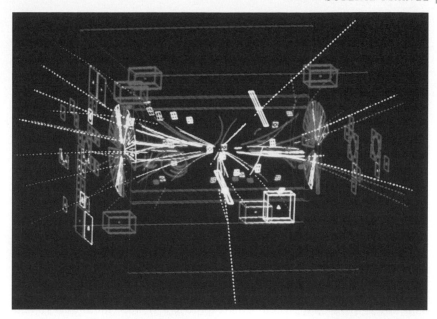

Figure 4.4 The first *Z* particle recorded by UA1 on 30 April 1983. The *Z* disintegrates too quickly to be seen, but was identified through the tracks of the electron–positron pair produced in the decay process.

Source: CERN / UA1 Collaboration.

In 1983, both the UA1 and UA2 experimental collaborations discovered first the *W* and then the *Z* particle, and their masses were measured to be respectively about 85 and 95 times the proton mass, in perfect agreement with theoretical predictions. From these results, it was inferred that the range of the weak force is about 500 times smaller than the proton size. This is the basic reason for the observed weakness of the weak force. But the electroweak unification states that, at distances 500 times smaller than the proton size, there is no intrinsic difference between the strength of electromagnetic and weak forces. The discovery of the *W* and *Z* particles was the final confirmation of the electroweak theory.

Quantum chromodynamics

> Metaphorically, QCD stands to QED as an icosahedron stands to a triangle.
>
> Frank Wilczek[18]

---

[18] F. Wilczek, *Fantastic Realities*, World Scientific, Singapore 2006.

Before the discovery of quarks, the situation regarding the strong force appeared to be rather desperate, as summarized by the words of the Russian theoretical physicist Lev Landau in 1960: "It is well known that theoretical physics is at present almost helpless in dealing with the problem of strong interactions...By now the nullification of the theory is tacitly accepted even by theoretical physicists who profess to dispute it. This is evident...particularly from Dyson's assertion that the correct theory will not be found in the next hundred years."[19] Instead, the correct theory was found only 13 years later.

The discovery of quarks had imposed some order on the messy world of hadrons, but it had also opened new problems. As has been mentioned previously, the experiments by Friedman, Kendall, and Taylor at SLAC had provided a rather unexpected picture of the force among quarks. To use an analogy, the situation was the following. Imagine that quarks are joined together by elastic bands the length of the proton diameter, about 10^{-15} metres. If one tries to extract a quark out of the proton, the elastic band tends to pull it back inside. The further away the quark is, the stronger is the force that draws the quark back to its position inside the proton. This explains why individual quarks had never been identified in any experimental search: the elastic bands keep the quarks tightly together inside the proton. On the other hand, when quarks roam peacefully in the interior of the proton, they are at mutual distances smaller than the diameter of the proton. Thus the elastic bands are loose and they do not exert any force on the quarks. This explains why experiments have observed that quarks behave like free particles in their motion inside the proton. In scientific jargon, this picture was called *asymptotic freedom*, because quarks are nearly free from any binding force when they are very close together.

Asymptotic freedom was exactly the opposite of what was known about fundamental forces of nature. It was expected that any such force would become weaker as bodies are moved apart, which is the case for gravitational, electric, magnetic, and weak forces. Moreover, it was almost taken for granted that any quantum field theory described only forces growing weaker at larger distances.

David Gross (Nobel Prize 2004), of Princeton University, was also perfectly convinced of this and started to systematically analyse the problem with the intent of rigorously proving that asymptotic freedom is incompatible with any quantum field theory. In other words, he wanted to show that all forces in quantum field theory become weaker at larger distances. He made rapid progress and completed the programme with one single exception: gauge theory. After the success in explaining

[19] L.D. Landau, in *Theoretical Physics in the Twentieth Century; a Memorial Volume to Wolfgang Pauli*, Interscience, New York 1960.

the electroweak forces, the reputation of gauge theory had grown in the eyes of physicists, and it was really necessary to keep on working in order to examine that last case. At the end of 1972 Gross confronted the case of gauge theory together with his student Frank Wilczek (Nobel Prize 2004). Later the two learned that, at Harvard, Sidney Coleman had assigned to his student David Politzer (Nobel Prize 2004) a nearly identical problem.

The result of these studies was astonishing. Calculations showed that, in certain cases, gauge theories predicted exactly the phenomenon of asymptotic freedom. "For me the discovery of asymptotic freedom was totally unexpected", declared Gross. "Like an atheist who has just received a message from a burning bush, I became an immediate true believer."[20]

Once again gauge theory turned out to be the right answer. The strong force, like electromagnetic and weak forces, is correctly described by gauge theory. The carriers of the force (the analogues of the photon, W and Z in the electroweak theory) are called *gluons*. Gluons keep quarks inside hadrons like a form of "glue", but they have the peculiarity of propagating a force that becomes stronger as bodies are moved apart, in sharp contrast with the other known fundamental forces. And yet, the theory that describes the strong force is conceptually the same as the electroweak theory, in spite of the gross differences between the physical phenomena associated with them.

Electromagnetism is caused by the exchange of one particle: the photon. The weak force is due to the W and the Z. The strong force is mediated by gluons, which come in eight different species called *colours*. Quarks come in three different colours too. But you shouldn't expect detectors at the LHC to observe blue, red, and green quarks or multicoloured gluons. Colour is just a fictional word chosen by physicists to indicate an attribute of quarks and gluons, similar to the electric charge. You can think of colour as the charge of the strong force. Event reconstruction images from the LHC sometimes look like Jackson Pollock's pictures with many colourful tracks corresponding to various particles. However, don't be fooled: it is just a colour coding used by the digital reconstruction programme and it has nothing to do with the "colour" of quarks and gluons.

Physicists enjoy fanciful words and during a summer meeting at Aspen in 1973, Gell-Mann named the new theory of the strong force *Quantum Chromodynamics* (abbreviated as QCD). The reference is to the colour of quarks and gluons (from the Greek *chróma*, colour) and the acronym QCD underlies the conceptual similarity of the theory to QED.

[20] D. Gross, in *The Rise of the Standard Model*, ed. L. Hoddeson, L. Brown, M. Riordan, M. Dresden, Cambridge University Press, Cambridge 1997.

Some years earlier, Sheldon Glashow, John Iliopoulos, and Luciano Maiani were working hard trying to make sense of an unexpected property in the interaction of the Z particle with quarks. "Our collaboration soon developed a standard pattern," recalled Iliopoulos. "Each day one of us would have a new idea and invariably the other two would join to prove to him it was stupid."[21] But, eventually, they stumbled on an idea that was not at all stupid. The Eightfold Way explained the structure of hadrons in terms of three quarks, called *up, down* and *strange*. Glashow, Iliopoulos and Maiani understood that, in order to make sense of hadron interactions with the Z, a fourth kind of quark must exist, and they gave the name *charm* to this new hypothetical quark. "We called our construct the 'charmed quark', for we were fascinated and pleased by the symmetry it brought to the subnuclear world,"[22] declared Glashow.

"Ten days of November 1974 shook the world of physics. Something wonderful and almost unexpected was to see the light of day: a very discreetly charmed particle, a hadron so novel that it hardly looked like one,"[23] recounts Alvaro De Rújula. Two experimental groups in the USA, one at SLAC led by Burton Richter (Nobel Prize 1976) and one at Brookhaven National Laboratory led by Samuel Ting (Nobel Prize 1976), simultaneously discovered a particle, which received the awkward name J/ψ (by joining the two names used by the two different groups). After its significance was fully understood, this discovery had a double effect. It proved the existence of charm and it confirmed QCD, because the properties of the J/ψ could only be explained by asymptotic freedom.

The theoretical interpretation following the J/ψ discovery gave final confirmation to the hypothesis that the strong force is described in terms of quarks and QCD. Those events were later proudly referred to as the "November Revolution". "In a nutshell, the standard model arose from the ashes of the November Revolution, while its competitors died honorably on the battleground,"[24] in the words of De Rújula, one of the men who fought the battle in those glorious days.

Two more quarks were yet to come. Already in 1973 Makoto Kobayashi (Nobel Prize 2008) and Toshihide Maskawa (Nobel Prize 2008), extending a scheme first proposed by Nicola Cabibbo, predicted the existence of the *bottom* quark (sometimes called *beauty*) and of the *top*

---

[21] J. Iliopoulos, *What a Fourth Quark Can Do,* in *The Rise of the Standard Model,* ed. L. Hoddeson, L. Brown, M. Riordan, M. Dresden, Cambridge University Press, Cambridge 1997.

[22] S. Glashow, *The Hunting of the Quark,* The New York Times Magazine, 18 July 1976.

[23] A. De Rújula, in *Fifty Years of Yang–Mills Theory,* ed. G. 't Hooft, World Scientific, Singapore 2005.

[24] A. De Rújula, *ibid.*

quark. Experimental confirmation followed, although the top quark was discovered only in 1995 at the Tevatron, the proton–antiproton collider built at Fermilab.

The standard model

> All models are wrong but some are useful.
>
> George Box and Norman Draper[25]

So all the pieces of the puzzle have finally fallen into place and the result is an authentic Sublime Marvel of scientific achievement. This Sublime Marvel describes electromagnetic, weak, and strong forces and all known forms of matter in terms of a single principle – that of gauge theories. It is just astounding that the complexity of nature is the result of a single underlying principle, and it is equally astounding that human intellect was able to identify such a principle.

The Sublime Marvel can be summarized in a single equation or, though much less precisely, in a few sentences. Matter is formed by quarks and leptons, which fit a repetitive structure. Electromagnetic, weak and strong forces are all described by the same theoretical framework, in which forces are transmitted by particle carriers: photon, *W, Z,* and gluons. The Sublime Marvel is today referred to by the acronym SM, which actually stands for its true, though exceedingly modest, name: the *Standard Model.*

Many high-energy accelerators have contributed to test the Standard Model, most notably the Tevatron at Fermilab, LEP at CERN, HERA at the German laboratory of DESY, and SLC at SLAC. Experiments at lower energy facilities have also performed essential tests. In all cases, the theoretical predictions extracted from the Standard Model have been fully confirmed, at an astonishing level of precision. It is rare in science to find a theory so conceptually simple and yet so vast in its domain of application, so fundamental and yet so well experimentally tested as the Standard Model.

Is the Standard Model the end of the story? In spite of the experimental success and the theoretical elegance of the Standard Model, the answer is a firm no. There are too many open questions that have not yet found an answer. Are quarks and leptons the fundamental entities of nature or is there another layer in the structure of matter? Why are quarks and leptons sequentially repeated three times? These are very puzzling questions, because any form of matter and any phenomenon that we normally observe are well explained by the existence of the

---

[25] G.E.P. Box and N.R. Draper, *Empirical Model-Building and Response Surfaces,* Wiley, New York 1987.

	First generation	Second generation	Third generation	Electric charge
Quarks	Up (0.003 GeV)	Charm (1.3 GeV)	Top (173 GeV)	$\frac{2}{3}$
	Down (0.005 GeV)	Strange (0.1 GeV)	Bottom (4.2 GeV)	$-\frac{1}{3}$
Leptons	Electron neutrino ($<10^{-9}$ GeV)	Muon neutrino ($<10^{-9}$ GeV)	Tau neutrino ($<10^{-9}$ GeV)	0
	Electron (0.0005 GeV)	Muon (0.1 GeV)	Tau (1.8 GeV)	-1

Gauge particles (force carriers)			Electric charge
Strong force		Gluons (zero mass)	0
Weak force		W (80 GeV)	1
		Z (91 GeV)	0
Electromagnetic force		Photon (zero mass)	0

Figure 4.5 The particle content of the Standard Model. The numbers in parentheses are the particle masses expressed in GeV, a unit of mass commonly used in particle physics.

fundamental forces and by the up and down quarks, the electron and one neutrino. These particles are called the *first generation* of quarks and leptons. All other quarks and leptons appear to be completely superfluous for our world, save for appearing in some particle-physics experiments and cosmic-ray interactions. And yet, the Standard Model replicates three times the simplest structure of the first generation of quarks and leptons. These repetitions, called the *three generations*, contain particles that are perfectly identical, generation by generation, except for one property: their masses. Up, charm and top quarks are completely indistinguishable, save for their masses. The same happens for down, strange and bottom quarks; for the electron, muon, and tau; for the three neutrinos. The problem of understanding the origin of this structure is just an expanded version of the question raised by Rabi regarding the muon: "Who ordered that?" We have not yet found the answer.

How does gravity fit into the scheme? Gravity, the most familiar of all forces, is actually ignored by the Standard Model. For practical purposes, this is not a problem. The gravitational force between two electrons is 10^{43} times weaker than the corresponding electrostatic force. This means that it is absolutely justified to neglect gravity with respect to electromagnetism in the interaction between two electrons. This approximation is as accurate as measuring the size of the universe neglecting the length of a simple nucleus.

The extreme feebleness of gravity in the particle world may surprise us for we are used to thinking of gravity as a powerful force. But in the macroscopic world gravity is so intense only because its attractive pull is a cumulative effect summed over a huge number of matter constituents. On the other hand, the charge neutrality of atoms forming matter largely screens the electrostatic force, which then becomes usually less important than gravity at large distances. But, even if gravity is essentially irrelevant in particle-physics experiments, its embedding in a complete picture of fundamental forces is a crucial open problem in theoretical physics.

All these are fundamental questions that need to be addressed. But there is an even more urgent issue. The Standard Model – as defined by the simple combination of quarks, leptons, and force messengers – is actually incomplete. There is still one missing element: how do elementary particles obtain their masses? To understand how nature solves this formidable question, we need a formidable accelerator: we need the LHC.

PART TWO
THE STARSHIP OF ZEPTOSPACE

5

Stairway to Heaven

———◦◦◦◦———

'Cause you know sometimes words have two meanings,... when all
are one and one is all.

Led Zeppelin[1]

The Unification of Science

As the kabbalist said to the hot-dog vendor, 'Make me one with
everything.'

Rabbi Lawrence Kushner[2]

At the turn of the 20th century, it was a common belief among scientists
that physics had completed its task of discovering the laws of nature.
Maxwell wrote in 1871: "The opinion seems to have got abroad that, in
a few years, all the great physical constants will have been approximately estimated, and that the only occupation which will then be left to
men of science will be to carry on these measurements to another place
of decimals."[3] Kelvin reiterated in 1900: "There is nothing new to be
discovered in physics. All that remains is more and more precise
measurement,"[4] and Michelson added in 1903: "The most important
fundamental laws and facts of physical science have all been discovered, and these are now so firmly established that the possibility of their
ever being supplemented in consequence of new discoveries is exceedingly remote."[5] Only a few years later, relativity and quantum mechanics

---

[1] Led Zeppelin, *Stairway to Heaven,* Atlantic 1971; lyrics by Robert Plant.

[2] L. Kushner, *Annals of the New York Academy of Sciences*, 950, 215 (2001).

[3] J.C. Maxwell, *The Scientific Papers,* ed. W.D. Niven, Dover, New York 1965.

[4] It is often said that Kelvin pronounced these words at the 1900 Annual Meeting of
the British Association for the Advancement of Science. Unfortunately, it is not possible
to ascertain the authenticity of this quote for, to the best of my knowledge, there is no
written record of his speech.

[5] A.A. Michelson, *Light Waves and Their Uses*, The University of Chicago Press,
Chicago 1903.

would overturn almost all known principles of classical physics and revolutionize science.

Even great physicists sometimes make wrong predictions. It is well known that Kelvin at first dismissed X-rays as a hoax. Moreover, in 1896 he wrote: "I have not the smallest molecule of faith in aerial navigation other than ballooning or of expectation of good results from any of the trials we hear of."[6] Only seven years later, the Wright brothers took off over Kitty Hawk, in North Carolina. Actually, Lord Kelvin had even earlier experiences with wrong predictions. Legend has it that, while studying at the University of Cambridge, he was so firmly convinced to be named *Senior Wrangler* (the highest-scoring student) at the famous Mathematical Tripos exams that he asked his servant to run to the Senate House and check who had been named *Second Wrangler* (the second highest-scoring student). The servant came back and told him: "You, Sir."[7]

But, admittedly, it was reasonable for physicists at the end of the 19th century to believe that physics knowledge was essentially complete. Every phenomenon could be explained in terms of Newton's mechanics, Maxwell's electromagnetism, thermodynamics, optics, or fluid mechanics. The real revolution of 20th century physics was to show that all these islands of knowledge are actually the emerging tips of a unique and fundamental conceptual structure that can simultaneously explain all natural phenomena.

The idea of unification in science is quite old. The first brilliant example is the understanding that the same force, gravity, is responsible both for the motion of celestial bodies and for falling objects on earth. Even before the work by Newton, Galileo guessed the logical connection between these very different phenomena, with remarkable farsightedness. Fifty-five years before the publication of Newton's *Principia*, Galileo made Salviati – his alter ego in the *Dialogue* – say: "But if this author knows by which principle other world bodies are moved in rotation, as they certainly are moved, then I say that that which makes the earth move is a thing similar to whatever moves Mars and Jupiter, and which he believes also moves the stellar sphere. If he advises me as

---

[6] Letter to Major Baden-Powell, 8 December 1896; reprinted in J.L. Pritchard, *Journal of the Royal Aeronautical Society* 60, 9 (1956).

[7] The 1845 title of Senior Wrangler went instead to Stephen Parkinson, who later became a mathematician at the University of Cambridge, although he never achieved the fame of Kelvin. It is curious to note that both James Clerk Maxwell and Joseph John Thomson were named Second Wrangler in their respective years as well. Maxwell lost the title to Edward Routh, who subsequently became a mathematician known especially as a coach of Senior Wranglers, because for 22 consecutive years one of his students was named Senior Wrangler. Thomson was beaten by Joseph Larmor, who became a physicist known for his work in electrodynamics and thermodynamics.

to the motive power of one of these movable bodies, I promise I shall be able to tell him what makes the earth move. Moreover, I shall do the same if he can teach me what it is that moves earthly things downward."[8]

But it was Newton who fully elaborated this concept and, most importantly, put it into equations. He demonstrated that a single universal gravitational theory could explain both terrestrial and astronomical phenomena. Newton was firmly convinced that physics (or natural philosophy, as it was then called) should explain the complexity of nature in terms of simple fundamental forces. He tried to identify these forces using as a paradigm the laws of mechanics discovered by him, which, he believed, could be extended to any other phenomenon. In the introduction of the *Principia* he stated, in perfect tune with the approach of modern physics: "For the whole burden of philosophy seems to consist in this: from the phenomena of motions to investigate the forces of nature, and then from these forces to demonstrate the other phenomena. . . . I wish we could derive the rest of the phenomena of Nature by the same kind of reasoning from mechanical principles, for I am induced by many reasons to suspect that they may all depend upon certain forces."[9]

Maxwell's equations represented a gigantic step in this programme of unification, since electric and magnetic phenomena were explained by the same theory. The quest for a unified theory able to describe all forces continued with Einstein: "The supreme task of the physicist is to arrive at those universal elementary laws from which the cosmos can be built by pure deduction."[10] Since then, *unification* has become the leit-motif of fundamental physics. Unification means simplification and synthesis of the elements necessary to describe the physical laws but, above all, it means obtaining a new and deeper understanding of the principles of nature. Unification is not just an elegant intellectual exercise. Almost invariably, each step in the process of unification ushers in new unexpected discoveries: new phenomena predicted by the unified theory or new logical connections with other branches of scientific research. Who could have guessed that unification between electricity and magnetism would have led to the discovery that light is an electromagnetic wave?

The understanding that different natural phenomena do not follow independent laws, but have a common origin within a unitary framework, has been one of the greatest successes of 20th century science. As Einstein stated in anticipation of future developments in physics: "The

---

[8] G. Galilei, *Dialogue Concerning the Two Chief World Systems,* 1632.

[9] I. Newton, *Philosophiæ Naturalis Principia Mathematica,* 1687.

[10] A. Einstein, *Principles of Research,* address to the Physical Society of Berlin in 1918 for Max Planck's sixtieth birthday; reprinted in A. Einstein, *Ideas and Opinions,* Crown, New York 1954.

general laws on which the structure of theoretical physics is based claim to be valid for any natural phenomenon whatsoever. With them, it ought to be possible to arrive at the description, that is to say, the theory, of every natural process, including life, by means of pure deduction, if that process of deduction were not far beyond the capacity of the human intellect."[11]

The Standard Model is the highest level of unification that has been reached so far. Quantum fields, which manifest themselves as particles, are the fundamental ingredients of nature for both matter and force. But the Standard Model cannot be the final theory and the journey that physics has undertaken towards the ultimate laws of nature is not over. The LHC is the instrument needed to carry us further on this journey.

Jacob's ladder

> I want to know how God created this world. The rest are details.
>
> Albert Einstein[12]

The Book of Genesis narrates that Jacob, for fear of his brother Esau, left Beersheba and reached Haran. There he decided to spend the night and, resting his head on a stone, he fell asleep. "And he dreamed that there was a ladder set up on the earth, and the top of it reached to heaven; and behold, the angels of God were ascending and descending on it. And behold, the Lord stood above it."[13] Leaving aside any religious interpretation – of no concern to physics – Jacob's dream offers an insightful metaphor of nature's order.

The study of nature has taught us that many macroscopic phenomena can be understood in terms of microscopic entities. This process of reduction to more elementary components is repeated in successive steps. Matter is made of molecules, which are compounds of atoms; atoms are made of electrons orbiting around nuclei; nuclei are composed of protons and neutrons, which are made of quarks. Climbing Jacob's ladder is like moving towards smaller and smaller distances. At each step of Jacob's ladder we discover new fundamental entities, which change our way of looking at nature and provide the new ingredients for the most appropriate interpretation of the physical world.

Moreover, fundamental physics has revealed another important aspect of nature's order. The physical laws that rule the microscopic

---

[11] A. Einstein, *ibid.*

[12] A. Einstein, as quoted in *The Expanded Quotable Einstein,* ed. A. Calaprice, Princeton University Press, Princeton 2000.

[13] Genesis, 28: 12, 13.

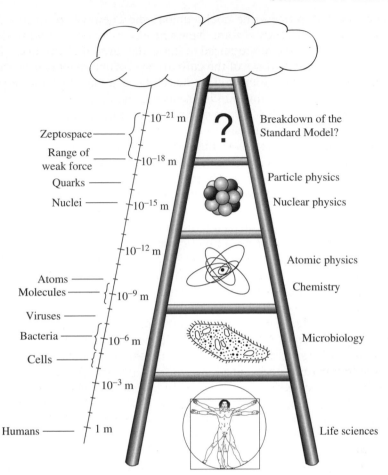

Figure 5.1 Jacob's ladder.

entities are simpler than those of the macroscopic world. By moving towards smaller distances, we discover that the variety and complexity of our world is merely disguising the simplicity of hidden fundamental laws. The apparent chaos of the macroscopic world is magically resolved into a neater order at each new step of Jacob's ladder. In a sense, this is similar to watching a picture on a computer screen. In its totality, the image presents a complexity of shapes and chromatic variations. But, as we zoom into the image, we realize that in reality it is formed by many pixels – tiny squares, each of the same size and of uniform colour.

The study of small distances has revealed yet another important point that can be explained with a metaphor. A good cook is able to extract the most tempting and delicious flavours out of the ingredients.

In the cooking process the cook exploits the chemical reactions that occur among molecules of food. Nevertheless, to be successful in their profession, cooks are not required to know the laws of chemistry. They just need to know the laws of the culinary art, which involve properties such as sweet, bitter, sour, and so on. Only to understand the deeper reasons for how a certain taste is produced, should they turn to chemistry. Neither is it necessary for a gourmet to understand chemistry in order to enjoy the cook's dishes.

An analogous situation, in a more scientific context, is found in thermodynamics. The properties of gases are perfectly well described by the laws of thermodynamics, which involve quantities such as temperature, pressure, and entropy. But by going one step further up Jacob's ladder, we discover that gases are made of molecules. The new interpretation provides a deeper understanding of the laws of thermodynamics, now given in terms of particle kinetic energy and statistical properties.

These examples show that each step of Jacob's ladder can be described by a coherent scientific theory, with no need of full knowledge of all the other steps. In simpler words, it means that we can formulate a consistent atomic theory without knowledge of nuclei; we can derive nuclear theory without knowledge of quarks; and so on to smaller and smaller distances. Each step of Jacob's ladder gives an adequate picture of nature at the appropriate distance scale. As we increase the distance scale at which we observe natural phenomena, we go from particle physics to atomic physics, chemistry, microbiology and life sciences. Each step is linked to the previous one, but it is governed by its own laws. At each step, nature presents new and interesting phenomena worthy of dedicated scientific investigation. In fact, these phenomena could not be properly described in terms of the elements of a different step of Jacob's ladder. For instance, the equations describing the motion of quarks are of little use in computing the macroscopic properties of gases.

Fundamental physics aims at climbing Jacob's ladder because the discovery of each new step provides a deeper understanding into the meaning of things. While one step describes *how* nature works, the next step explains *why* nature works that way. While the theoretical description of one step requires input parameters to be determined by measurements, in the next step some of these parameters become calculable quantities that can be predicted by the theory.

It may appear rather obvious that there is a separation in nature between phenomena at different distance scales. And yet, there is no inescapable logical necessity for why nature follows this behaviour. Certainly if this were not the case, the life of scientists (not to mention cooks) would be terribly difficult. The explanation of any physical phenomenon would be inextricably linked to the knowledge of every detail of nature at any distance scale. Newton could not have discovered the gravitational law without solving the equations that govern the motion of every single quark inside the moon. No physical process could be understood without knowledge of nature's behaviour at arbitrarily small distances. Fortunately, there is a Jacob's ladder in nature.

This separation of different distance scales in physics has a well-defined mathematical formulation, called the *effective theory*. An effective theory gives an approximate description of nature, which is obtained by truncating the effect of any physical process occurring at very small distances. The reason why such a truncation is possible is related to the property of locality of quantum field theory, already encountered in Chapter 4. Expressed in simpler words, an effective theory describes just one single step of Jacob's ladder.

The structure of Jacob's ladder is not a philosophical construct, but an empirical fact. Its existence is the reason why science could progress in the understanding of the particle world. Step by step, science advances building one effective theory on top of another. The primary motivation for continuing this process of discovery is that more fundamental physical laws describe each successive step. Concepts that seemed completely independent at one step became unified in a single entity at the next step. This result is what drives physicists to explore smaller and smaller distances, and to keep on ascending Jacob's ladder in the search of the universal laws of nature.

The LHC explores distances much smaller than those penetrated by any other previous experiment. But the great excitement about the LHC is not based on some vague notion of exploring new territories. As examined in the third part of this book, there are good reasons to believe that entering zeptospace corresponds to a jump onto a new step of Jacob's ladder, described by a new effective theory different from the Standard Model. If this is indeed the case, the LHC will ignite an intellectual revolution.

The understanding of nature's order brings about a spontaneous question: what stands above Jacob's ladder? Some physicists believe that a last and final step exists. They trust in an ultimate theory able to describe all forces and all forms of matter in a unified manner. At the top of the ladder we will find an exact theory, uniquely determined by logical consistency, in which there are no arbitrary fundamental constants or parameters.

Other physicists do not share this point of view and wonder: is there really a top of Jacob's ladder? Maybe there are actually an infinite number of steps so that nobody can ever reach the top. Like physicists at the end of the 19th century, we will be periodically convinced that we have discovered everything, until a new revolution of ideas dispels such beliefs and pushes physics towards new goals. And the process of discovery will have no end.

Reality could be different in yet another way: what if Jacob's ladder changes into something different at a certain height? Perhaps the physicists' views based on effective theories will fail beyond a certain distance. No truncation of small distances will be allowed any longer, no effective theories, but a new conception of our universe will emerge. This new conception ought to have a radically different mathematical formulation from the theories we know of today and would require a deep rethinking of the basic principles of nature.

Maybe one day we will know the answers to these lofty questions. For the moment, all we can do is to keep on ascending Jacob's ladder step by step like the angels of God.

Bigger and bigger microscopes

It is only in the microscope that our life looks so big.

Arthur Schopenhauer[14]

The hunt for the fundamental laws of nature guides us towards exploration of smaller distances. With the help of a microscope, we can discover that a living organism, at distances of tens of microns, is made of cells (a micron is equal to a millionth of a metre). We can try to increase further the magnification of the instrument, but no optical microscope can resolve images beyond some fraction of a micron, the size of the smallest bacteria. The reason is that light is an electromagnetic wave and it cannot be used to resolve any object smaller than its own wavelength. Visible light has wavelengths in the range between 380 and 750 nanometres (a nanometre is equal to a billionth of a metre). Any detail of the observed specimen much smaller than that will be necessarily blurred. It is not just a limitation of the specific optical instrument, but it is a consequence of an intrinsic property of light. It is like trying to measure the size of a gnat with a yardstick or to tighten the tiny screws of eyeglasses' frames with a car mechanic's screwdriver. A tool cannot be used to operate at distances much smaller than its characteristic size. In the same way, visible light has an intrinsic minimum length – its wavelength – and therefore it cannot resolve any distance smaller than a few hundreds of nanometres.

To explore nature beneath hundreds of nanometres we need probes with smaller wavelengths. Images of viruses and biomolecules are commonly obtained using electron microscopes. These instruments replace the beam of visible light used by ordinary optical microscopes with beams of electrons; and they replace the ordinary optical lenses with electromagnets. Electron microscopes can resolve distances up to few tenths of nanometres, the size of individual atoms, extending our

[14] A. Schopenhauer, *On the Vanity of Existence,* in *Studies in Pessimism,* Swan Sonnenschein, London 1893.

ability to explore the world of small distances. But this is still not sufficient to investigate nuclear and subnuclear matter. A more energetic
form of radiation, with smaller wavelength, is needed to delve into the
intimacy of subnuclear distances, where nature hides the secrets of the
fundamental laws of physics.

According to quantum mechanics there is a duality between waves
and particles. The meaning of this duality is that the real physical entity
is neither simply a wave nor a particle, but it has properties common to
both. Two concepts that appear at first sight as distinct – those of waves
and particles – are actually two expressions of the same essence in the
realm of quantum mechanics. For instance, we can interpret light as an
electromagnetic wave or as a beam of photons, and both descriptions are
correct. The same dual interpretation – à la Dr Jekyll and Mr Hyde – can
be extended from photons to any other elementary particle, as speculated in 1923 by the French physicist Luis de Broglie (1892–1987, Nobel
Prize 1929). When first proposed, the conjecture by de Broglie sounded
so preposterous that it came to be nicknamed "la Comédie Française".
Nevertheless, quantum mechanics brought many surprises, and electrons were indeed observed to show interference patterns typical of an
undulatory nature, confirming De Broglie's hypothesis.

The identification of particles and waves in one dual concept leads to
a relation between the energy of a particle and the wavelength of the
associated wave. The more energetic a particle is, the shorter its wavelength. Thus, according to quantum mechanics, a short-wavelength
radiation is equivalent to a beam of very energetic particles. Or, in other
words, to investigate matter at smaller and smaller distances, more and
more powerful particle accelerators are needed.

To explore a deep well, we can drop stones into it and from the delay
of the returning sound we can determine the depth of the well; from the
tonality of the sound, we can infer whether at the bottom there is water,
or soil, or something else. The same strategy is used to probe matter at
small distances, as illustrated by the experiments of Geiger, Marsden,
and Rutherford that led to the discovery of the atomic nucleus. Matter is
bombarded with highly energetic projectiles (which are equivalent to
short-wavelength radiation, according to quantum mechanics). The
higher the energy, the deeper the projectiles penetrate into matter. By
measuring the characteristics of the projectiles after their collisions with
the target, it is possible to extract information about what the projectiles
have encountered inside matter. The echoes of very energetic radiation
are interpreted and translated into an image of the microscopic world.

The logic of this approach is akin to the way that optical microscopes work. In optical microscopes, the projectiles are beams of light
that are reflected by the observed specimen and then perceived by our
eyes as an image. In modern experiments, the projectiles are highly
energetic particles while sophisticated electronic detectors are the "eyes"

used to observe the debris of the collisions and to reconstruct the "image" of nature at small distances.

There is another reason why the exploration of the particle world requires gigantic high-energy accelerators. To explain this point it is necessary to open a parenthesis. In 1905, a clerk from the Bern patent office, named Albert Einstein, introduced the most celebrated physics equation ever written, $E = mc^2$, later explaining that "mass is equivalent to an energy content."[15] The meaning of this equation is that mass (m) is a form of energy (E), in much the same way as heat or kinetic energy. The square of the speed of light (c^2) is the conversion factor between energy and mass, just as there is a conversion factor between euros and dollars, or between kilometres and miles. We can express a price in euros or dollars: the number will be different, but the value is the same. We can say that the distance between Geneva and Paris is equal to 404 km, or to 251 miles, without changing the meaning.

In the same way we can convert mass into units of energy. However, the conversion factor between energy and mass (in units of measurement familiar to us) is really huge, in contrast with the conversion factor between euro and dollar (at the moment, at least) or between kilometre and mile. For instance, a kilogram of matter, according to Einstein's equation, corresponds to an energy of about 20 megatons, which is the energy produced by the explosions of more than a thousand Hiroshima bombs. Or to put it another way, a kilogram of matter is equivalent to the energy generated by the engine of a Ferrari 430 Scuderia driving at full speed for about 8000 years. To produce that amount of energy, the Ferrari engine would require several million tonnes of fuel.

Because mass is conceptually equivalent to energy, physicists usually express particle masses in a unit of energy: the electronvolt (eV). One electronvolt corresponds to the energy gained by an electron when accelerated in vacuum by an electrostatic potential of one volt. Physicists often use multiples of the electronvolt: MeV (a million eV), GeV (a billion eV), and TeV (a thousand billion eV). Thus, it is common among scientists to say that the electron mass is equal to 0.51 MeV (rather than 9×10^{-31} kg) and that the proton mass is 0.94 GeV (rather than 2×10^{-27} kg), although MeV and GeV are actually units of energy and not of mass. These units of energy and mass will be used throughout the rest of the book.

How does Einstein's equation come in? Theories describing nature at small distances predict the existence of new particles much heavier than ordinary protons and neutrons. In order to prove or disprove the validity of these theories, physicists have to search for new particles by producing them in laboratory experiments. This can be done, according to Einstein's

---

[15] A. Einstein, *Jahrbuch der Radioaktivität und Elektronik*, 4, 411 (1907).

equation, by converting energy into mass. In particle collisions, large amounts of energy can be concentrated into small regions of space; this energy can materialize in the form of new particles. Therefore, sources of high-energy particles are needed both to probe the inner properties of matter through short-wavelength radiation and to discover new kinds of particles by converting the colliding beams into unknown forms of matter.

In his experiments, Rutherford used beams of alpha particles produced by natural radioactivity. This allowed him to probe distances much smaller than what can be seen with visible light. However, alpha radiation cannot exceed a maximum energy characteristic of the radio-active material and, like visible light, cannot be used to probe arbitrarily small distances. The associated wavelength of alpha radiation – its measuring rod – ranges in the millionths of nanometres, sufficient to discover the atomic nucleus, but not small enough to further explore the world of particles.

It soon became clear that the exploration of the subnuclear structure needed an artificial way to accelerate particles beyond the energies produced by natural radioactivity. Rutherford himself recognized this need in 1927: "It has long been my ambition to have available for study a copious supply of atoms and electrons which have an individual energy far transcending that of the α- and β-particles from radioactive bodies. I am hopeful that I may yet have my wish fulfilled, but it is obvious that many experimental difficulties will have to be surmounted before this can be realized, even on a laboratory scale."[16] The LHC certainly fulfils Rutherford's wish but, as he correctly predicted, the path to it had many difficulties to be surmounted. It took all the ingenuity of generations of physicists and engineers to complete the journey that started with the first electrostatic accelerators developed during the 1930s at the Caven-dish Laboratory by John Cockcroft (1897–1967, Nobel Prize 1951) and Ernest Walton (1903–1995, Nobel Prize 1951) and that now has reached the construction of the most powerful accelerator ever – the LHC.

The many uses of accelerators

> The production of too many useful things results in too many useless people.
>
> Karl Marx[17]

---

[16] E. Rutherford, *Address of the President at the Anniversary Meeting, November 30, 1927*, Proceedings of the Royal Society of London, Vol. 102, No. 717, p. 239 (1928).

[17] K. Marx, *Human Requirements and Division of Labour Under the Rule of Private Property*, in *Economic and Philosophical Manuscripts of 1844*.

Research in accelerators not only delivered some of the most powerful tools to investigate the particle world, but also led to unexpected spin-offs useful for practical purposes. Less than one per cent of existing accelerators are high-energy devices used for particle-physics research. The vast majority are small accelerators operating at low energies in hospitals around the world to produce radioactive isotopes or radiation beams for cancer therapy.

Ernest Orlando Lawrence (1901–1958, Nobel Prize 1938) was one of the first scientists to realize the medical applications of accelerator research and one of its most active advocates. He is best known among physicists as the inventor of the cyclotron, the first circular accelerator. He is also famous for having transformed the world of experimental particle physics by organizing large research teams and by raising substantial financial support from government and private funds. This way of managing science was unusual at the time and not easy to achieve, especially so soon after the Great Depression, but it was necessary to meet the challenges posed by the exploration of the particle world. As one of his collaborators later remarked, "the trade of a 'cyclo-troneer' is one which has experienced no depression."[18] Lawrence ran the laboratory in Berkeley (now called Lawrence Berkeley National Laboratory to honour his legacy) with a mixture of passion, sternness and camaraderie. An anecdote tells us that he once burst into an office where he saw a man with a telephone receiver in his hands and the feet leisurely stretched over the desk. "You are fired!" he shouted, in his usual impulsive manner. The man looked at him half-puzzled, half-defiantly and replied: "You can't fire me; I work for the phone company."[19]

At the cyclotron, Lawrence regularly produced radioactive isotopes, delivering them for free to hospitals and to research institutions. In 1937, his mother was diagnosed with a form of inoperable cancer. In collaboration with his brother John, a medical doctor at Yale, Lawrence treated her with X-rays and neutron beams. The therapy was successful, although a retrospective review of the case showed that the diagnosis was probably wrong. Perhaps the best we can say is that Lawrence's mother survived in spite of the treatment. But accelerator-based biomedical research has made great progress since then. Hadron beams (made up of protons or ions) are now considered the most promising technique, since they deposit almost all of their energy at a particular depth within

---

[18] F.N.D. Kurie, *Journal of Applied Physics,* 9, 691 (1938), quoted in J.L. Heilbron and R.W. Seidel, *Lawrence and His Laboratory. A History of the Lawrence Berkeley Laboratory: Volume I,* University of California Press, Berkeley 1990.

[19] A. Sessler and E. Wilson, *Engines of Discovery,* World Scientific, Singapore 2007.

the body, targeting cancerous cells and thus limiting damage to the patient's healthy tissues and sensitive organs. On the other hand, traditional X-rays lose most of their energy close to the skin surface, causing more damage to healthy cells than hadron-based treatments. Several facilities of hadron-therapy for curing tumour diseases are under construction in Europe and Japan.

Another extraordinary application of accelerator research is *synchrotron radiation*. When a beam of electrons is bent by a magnetic field, as in circular accelerators, it emits an electromagnetic wave, called synchrotron radiation. The physical process of emission is akin to the transmission of radio waves from an antenna. But the synchrotron radiation is concentrated in a narrow cone tangential to the electron beam, and it has a broad spectrum of frequencies, which depends on the energy of the electron beam. Since emission can occur also in the range of visible frequencies, synchrotron radiation can be seen by the naked eye or, as a safety precaution, photographed with a normal camera. Accelerator physicists first regarded synchrotron radiation as a nuisance because it degrades the energy of the electron beam: precious energy used to accelerate electrons for particle-physics experiments was instead dissipated into useless radiation. However, it was soon realized that synchrotron radiation offered a unique source of X-rays. Indeed its intensity is far greater then any previously known X-ray device. Modern synchrotron facilities can produce X-ray beams a million times more intense than those produced by ordinary medical equipment used in hospitals.

Synchrotron radiation sources are employed in much the same way as light is used in optical microscopes. But synchrotron light allows the observation of objects with a resolution in the range of nanometres. It has become an indispensable tool in a large variety of research fields. Being able to image the atomic and crystal structure, synchrotron light is used in nanotechnology to produce, for instance, microelectronic circuits or microsurgical instruments. In applied sciences, it has fostered the development of new materials. Synchrotron techniques are also used to detect material stress, as for instance wear in aircraft turbines. In biology, medicine, and pharmacology it has led to several breakthroughs, allowing direct studies of proteins and various biomolecules. It allows non-destructive tests of medical samples sensitive to the presence of microscopic quantities of substances, indistinguishable with other methods. There is growing interest in the use of synchrotron light sources in art restoration and archaeology. The Dead Sea Scrolls have been analysed using synchrotron light to obtain information on their textile fibres and on the pigments of the dye, which will also allow for precise dating. Non-destructive tests of the chemical composition of paintings help in understanding the causes of their deterioration and in choosing the appropriate restoration technique. By means of synchrotron light, for instance, it has been possible to discover the chemical explanation for the mysterious darkening of the crimson pigment in the 2000-year-old frescos at Pompeii.

Many new synchrotron-light sources are constructed or planned around the world. An interesting example, to appreciate the social role of science as well, is the SESAME project (Synchrotron light for Experimental Science and Applications in the Middle East). Its peculiarity is that it is built in Jordan by a scientific collaboration involving Israel, Iran, Pakistan, and several Arab countries including the Palestinian Authority. The project is in much the same spirit as the foundation of CERN, which has brought together scientists from nations that, just a decade before, were foes in the bloodiest war ever fought. Recently at CERN, some Israeli and Palestinian students organized a party where their respective flags were joined together by a banner saying "Because things can be different" and the word "Peace" in English, Hebrew, and Arabic.

A rich load of technological spin-offs has come not only from accelerators, but also from the research and development of particle detectors. Positron emission tomography (PET) is a medical imaging technique able to produce three-dimensional pictures of functional processes in the body. Certain positron-emitting substances are attached to biologically active molecules and then introduced into the body. The annihilation of positrons produces gamma rays, which are detected by scanners. Computers read the digital information and reconstruct a full image.

Commercial digital X-ray imaging has developed from research on detection of particle tracks in collider experiments. This technique allows real-time analysis of X-ray images, essential in many medical procedures. It also greatly reduces the required radiation doses with respect to ordinary photographic X-ray images, limiting the risk of tissue damage. The same technology is routinely used in on-line scanning of luggage, and even freight containers and trucks.

Silicon microstrips developed to detect the passage of electrically charged particles are now used to model the process in which human vision works. Under study is the possibility of using this technology to produce artificial eye retinas that could restore normal vision functions in some blind individuals.

Particle detector techniques are also having both useful and amusing applications in unexpected sectors. Research on the inner detector of an LHC experiment has been recently applied to optically measuring with great precision the grooves in old music discs. Historical recordings that could soon be lost because of deterioration are digitized with refined accuracy without any risk of damaging the original discs, because there is no physical contact with the material.

Colliders

> Necessity, who is the mother of invention.
>
> Plato[20]

All the unexpected benefits notwithstanding, the main role of research on accelerators and detectors is to progress in the exploration of smaller distances. Accelerators create high-energy particle beams that are directed to hit a target composed of a thin layer of matter. But, rather than using a single beam aimed at a stationary target, a much deeper probe of matter can be achieved by a *collider*, where two particle beams travelling in opposite directions are brought to a head-on collision. The greater effectiveness of colliders is fairly obvious: just compare the consequences of a fast car hitting a parked vehicle to a head-on crash between two cars driving at full speed towards each other. Indeed, if one of the LHC beams were simply directed against a target of stationary protons, the energy released in the collision would be many thousands of times less than what is actually achieved by the collider, and the capability of the LHC to travel into the depth of zeptospace would be nil.

Although the advantages of colliders have long been clear, the first prototypes were built only in the early 1960s. The great challenge of colliders was to create particle beams intense enough and focused enough to have a reasonable probability of producing direct head-on collisions. Particles are indeed so small that typically two opposing beams cross each other unscathed, like two intersecting rays of light. Only when the beam is highly squeezed and intense, can particles from one beam have a reasonable chance of hitting incoming particles from the other beam. A collider is like a highway system built by a deranged civil engineer. The lanes of this peculiar highway are much wider than the size of cars, but opposite lanes sometimes cross each other, without any warning or any traffic light. However, in spite of the recklessness of our civil engineer, head-on car crashes rarely happen, because the lanes are so wide that cars usually miss each other. Accidents occur more often during rush hour, when traffic becomes very heavy. Physicists need to achieve conditions of heavy traffic and of frequent crashes to produce the bursts of energy required to create new particles.

Roughly speaking, the performance of a collider is determined by three characteristic properties. The first is the kind of particles accelerated in the beam. The LHC mostly operates with protons although, for

---

[20] Plato, *The Republic*.

shorter running periods, it will accelerate heavy nuclei. Previous colliders have also used oppositely circulating proton–antiproton, electron–positron, electron–proton, or positron–proton beams.

The second parameter is the *energy* of the accelerated particles inside the beam. The higher the energy, the more violent is the collision. Higher energy beams correspond to radiation of smaller wavelength, thus providing a deeper probe inside matter. The LHC is designed to reach proton energies of 7 TeV in each beam (TeV is equal to one thousand billion electronvolts), which corresponds to a wavelength of less than 30 zeptometres. Thus the LHC energy is well suited for a direct exploration of zeptospace. The LHC energy is the highest ever reached by an accelerator although particles in the cosmos are routinely accelerated up to much higher energies in violent astrophysical environments. Fluxes of these particles travel through the cosmos, and our atmosphere is constantly bombarded with a rain of particles, which can be millions of times more energetic than those produced at the LHC. These are the cosmic rays, which were used for the early discoveries of particle physics. Unfortunately cosmic rays do not come in neatly arranged intense beams to be used for controlled experiments, and cannot compete with the LHC in performing a systematic exploration of zeptospace.

The third parameter that defines the properties of a collider is its *luminosity*. Luminosity is a precisely defined quantity which, roughly speaking, gives a measure of the intensity of the beams and therefore of the frequency of particle collisions. Energy without luminosity is of little use to particle physicists. If only a handful of cars are using the highway of the deranged civil engineer, accidents will be unlikely. Cars travelling at higher speed produce more spectacular crashes, but heavy traffic is needed to produce a sufficiently large number of accidents. At the LHC not only energy, but luminosity too is particularly high. This is very important to observe the rare and unfamiliar phenomena that we expect to occur in zeptospace. However, as we will see later, high luminosity also imposes formidable technological challenges.

6
The Lord of the Rings

Not all those who wander are lost.

John Tolkien[1]

Birth of a giant

Politics is not an exact science.

Otto von Bismarck[2]

The first ideas and feasibility studies for the LHC date from the beginning of the 1980s, but the meeting held in Lausanne in 1984 is usually considered the event that marks the birth of the project. In that meeting the proponents of the LHC addressed the challenges of the construction and outlined the characteristics of the machine. The original 1984 plan foresaw proton beams with energies up to 10 TeV, instead of the 7 TeV of the final LHC design, but a lower luminosity.

The early 1980s were a very exciting time for CERN. In 1983, the two carriers of the weak force – the W and Z particles – were discovered at the proton–antiproton collider. Around the same time, construction of LEP had started at CERN. LEP was an accelerator colliding a beam of electrons against one of positrons (the antiparticles of the electron) at an energy appropriate to studying in great detail the properties of the W and the Z. Electron–positron colliders are ideal machines for obtaining precision measurements because electrons – in contrast to protons, which are complicated composites made up of quarks and gluons – are truly elementary particles, as far as we know. This property allows a

---

[1] J.R.R. Tolkien, *The Fellowship of the Ring (The Lord of the Rings, Volume one)*, George Allen & Unwin, London 1954.

[2] O. von Bismark, speech to the Prussian Upper House, 18 December 1863, as quoted in *The Oxford Dictionary of Quotations*, Oxford University Press, Oxford 1941.

clear and plain interpretation of data from electron–positron colliders. Experiments at LEP were able to make spectacularly precise measurements, fully confirming the validity of the Standard Model and establishing it as the triumphant sovereign of the particle world.

Circular electron–positron colliders, like LEP, have the great disadvantage that energy cannot be increased at will because of a fundamental limitation: synchrotron radiation, the same phenomenon that leads to the fascinating and useful applications described in Chapter 5. However, from the point of view of particle physics, synchrotron radiation is simply a vexing parasite of colliders, because it takes away energy from the particle beam. At LEP, the electron beam lost about 3 per cent of its energy to synchrotron light at every turn. This is still acceptable, but the amount of emitted radiation grows very rapidly as the beam energy is increased. Upgrading the LEP energy from 100 GeV to 1 TeV would bring up radiation loss by a factor of 10 000. Energy from a huge number of power plants would not be enough to replenish the particle beam from the effect of energy loss in synchrotron radiation. For this reason, constructing a circular electron–positron collider much more powerful than LEP is unrealistic. The future of electron–positron machines can only be realized with linear colliders, which accelerate particles along straight trajectories. But linear colliders come with their own challenges, for electrons and positrons have to be accelerated within a relatively short distance, and the beams, after the collisions, cannot be reused, as they are in circular accelerators. At present there is a vigorous ongoing research programme on future linear colliders.

Scientists at CERN were well aware of the limiting factor of synchrotron radiation and they had the foresight to construct the 27 km long LEP tunnel wide enough to fit the necessary equipment for a proton collider, a possible successor of LEP. Indeed, proton colliders have almost no problem with synchrotron radiation. Because protons are heavier than electrons, a bent proton beam emits ten thousand billion times less radiation than an electron beam, under the same conditions. At the LHC, synchrotron emission is very limited (although it can be photographed with an ordinary camera). It amounts to only 3.6 kilowatt per beam – which is about the same energy consumption of a large kitchen oven. Still, its effect had to be carefully taken into account in the LHC design.

While CERN was enjoying the success of the W and Z discoveries, getting busy with LEP and planning the future LHC, scientists in the USA understood that they had to invigorate their particle physics programme so as not to lose ground. Since the end of World War II, the USA had been setting the pace for the main developments in the field. The US government was generously supporting research in particle physics for at least two reasons. The first was the recognition of the contribution to the war given by the Manhattan Project. The second was the appreciation that fundamental scientific research can fuel progress for society and drive economic growth. On the other hand, in the post-war period, single European countries, in spite of their prestigious universities, did not have the resources to support a vigorous programme in particle physics.

In the difficult years after World War II, Pierre Auger and Louis de Broglie in France, Edoardo Amaldi in Italy, Niels Bohr in Denmark, and several others started a visionary project for a common European laboratory devoted to fundamental research. In 1950, at a UNESCO Conference in Florence, the American Isidor Rabi drafted a resolution recommending the creation of a European Laboratory. Several US physicists, who had been trained or had worked in Europe before the war, were instrumental in promoting the idea of a European physics laboratory. Robert Oppenheimer said that European nations "would no longer be able to remain scientific leaders unless they pooled their money and talent" and that "it would be basically unhealthy if Europe's physicists had to go to the United States or the Soviet Union to conduct their research."[3]

Two years after the UNESCO resolution a provisional committee was set up, the *Conseil Européen pour la Recherche Nucléaire* (CERN), with the mandate to establish the new organization. On 1 July 1953 a Convention was signed in Paris by twelve countries creating the *European Organization for Nuclear Research*. Physicists, with their usual sense of logic and order, kept on calling the organization CERN, even though the original *Conseil* (the "C" in the acronym) was soon dissolved after its mandate was over. Furthermore, by now only a small fraction of the activity of the organization is devoted to nuclear research (the "N" in the acronym), while the vast majority deals with particle physics.

The twelve founding member states of CERN were: Belgium, Denmark, France, Germany, Greece, Italy, the Netherlands, Norway, Sweden, Switzerland, the UK, and Yugoslavia (which withdrew in 1961). At present, the number of member states has grown to twenty, after the admission of Austria (1959), Spain (1961, which however withdrew from 1969 to 1983), Portugal (1985), Finland (1991), Poland (1991), Hungary (1992), the Czech Republic (1993), the Slovak Republic (1993), and Bulgaria (1999).

According to the Convention, "the Organization shall provide for collaboration among European States in nuclear research of a pure scientific and fundamental character,... shall have no concern with work for military requirements and the results of its experimental and theoretical work shall be published or otherwise made

---

[3] R. Oppenheimer, as quoted in F. de Rose, *Nature* 455, 175, 2008.

Figure 6.1 A view of the main CERN site at the border between Switzerland and France.
Source: CERN.

generally available."[4] The goal was to foster fundamental research and encourage young physicists to remain in, or return to Europe. It is not easy today to appreciate the difficulties in establishing CERN and in bringing together nations that had been recently divided by war and people who had been educated to hatred by decades of propaganda. But the visionary project worked, and the discovery of the W represented the final step in the long path that took European physics to compete at the same level with American research.

But some policy makers in the USA saw this comeback of European particle physics as a sign of decline in American science. In 1983, soon after the discovery of the Z particle (often referred to as Z-zero, to specify that it has no electric charge) the *New York Times* published an editorial under the headline "Europe 3, U.S. Not Even Z-Zero"[5] emphasizing the need for the USA to regain its leadership in the field. Abandoning the construction of a previously planned project at the Brookhaven National Laboratory, the USA channelled most of the resources for particle physics into a gigantic new accelerator, the SSC (Superconducting Super Collider). The design foresaw an 87 km long underground

---

[4] *Convention for the Establishment of a European Organization for Nuclear Research*, Paris, 1 July 1953, Article II.
[5] The New York Times, 6 June 1983.

ring to host the machine that would accelerate and collide proton beams of 20 TeV, almost three times more powerful than what is achieved at the LHC.

The competition seemed unbearable. A proportion of the European physicists expressed the opinion that the LHC project should be stopped and that collaboration with the Americans on the SSC should be pursued. But there were also arguments against this point of view.

The cost for the LHC was much less than that of the SSC, whose price tag was about 5 billion dollars in 1986. Any significant European contribution would have nearly amounted to the full construction cost of the LHC. The LHC was much cheaper not only because of its more limited size and capacity, but also because it could exploit much of the infrastructure already existing at CERN, such as the LEP tunnel and the injector complex that carries out the preliminary stages of proton acceleration. On the other hand, the USA – for a mixture of reasons that included, among other things, politics, economics and geology – decided to build the SSC in a brand new laboratory located near Waxahachie, Texas, where only empty fields previously existed.

Another argument in favour of pursuing the LHC project was that experiments could perform an exciting physics programme, even in the presence of the more powerful SSC. This was especially true because of the high design luminosity of the LHC – that is, its extremely intense proton beams – which could compensate, although only partially, for the lower energy. Versatility was also an asset, because the LHC could be designed to operate with protons, nuclei, and even to collide protons against a LEP electron beam – an option that was later abandoned.

At the beginning of 1987, President Reagan approved the SSC project and the following year construction of the laboratory site and of the accelerator tunnel started in Waxahachie. On the other side of the Atlantic, a planning committee chaired by Carlo Rubbia recommended development of magnet technology for the LHC. Rubbia, who became CERN's director general in 1989, had always been a stubbornly enthusiastic and vibrant promoter of the project during the years of planning, research, and development.

The SSC had to obtain a vote of approval from Congress every year while costs were rising. In 1990, the cost had reached about 8 billion dollars, and Congress limited its contribution, requesting that special funds should be obtained from the state of Texas and from foreign contributors. However, it was not easy to procure financial support from abroad, after the SSC had been portrayed as a national project and especially after early expressions of interest for collaboration from Japan had been shunned. In the meantime, estimates of the cost had mounted to 11 billion dollars. Notwithstanding the strong scientific motivations of the project put forward by the physicists involved, and in spite of having already invested 2 billion dollars, Congress cancelled the SSC in October 1993, under the then new Clinton administration.

Much has been said and written about the termination of the SSC. Undoubtedly, at that moment, humanity missed a great opportunity to explore nature and expand technological and intellectual knowledge. Many causes have been blamed for this defeat: rising costs, project mismanagement, budgetary restrictions, lack of interest in fundamental physics, change in presidential administration, and support for alternative scientific projects. Whatever it was, it had a long-lasting, devastating effect on the particle-physics community around the world, hitting especially hard in the USA.

Things did not go very smoothly in Europe either, however. Admittedly, CERN offered a more stable funding system with a fixed budget where member states contribute with a fraction of their gross national products. The participating nations have always been very supportive of CERN's scientific mission and have often underlined their strong commitment to the LHC. Problems nevertheless were looming. Germany had already obtained a reduction of its contributions because of reunification costs, and both Germany and the UK were determined to veto any increase of the CERN budget related to the construction of the LHC. Under these conditions, the project was in danger. In the meantime, on the scientific side, research on the accelerator facility was progressing well. In November 1993 an external committee, chaired by Robert Aymar, reviewed the project, concluding that the technology was feasible, costs had been optimized, and safety assured.

Christopher Llewellyn Smith, the British theoretical physicist who succeeded Rubbia as CERN's director general at the beginning of 1994, started intense negotiations to obtain the approval for the LHC from the CERN Council, the governing body composed of the representatives from the various member states. Llewellyn Smith was facing the problem of fitting the LHC project into a tight CERN budget which, taking into account the effect of inflation, was effectively reduced. He started a process of revision of LHC costs and, at the same time, reduction to a bare minimum of any research activity unrelated to LHC and LEP. Then, with special financial help from the two host states – Switzerland and France, which are considered to receive additional economic benefits from the presence of CERN – an agreement was finally reached. The CERN Council would give the green light to the LHC project, but with a two-phase construction. During the first phase, only two-thirds of the dipole magnets would be installed. The rest would come only at a later stage. This meant that, during the first phase, the LHC could operate, but only at reduced energies. This solution of a degraded machine was not ideal from the physics point of view. The two-phase operation would even increase total costs, but it would allow CERN to postpone expenditure, thus remaining within its yearly budget. With this proviso, the CERN Council unanimously approved the LHC on 16 December 1994. The most ambitious and challenging scientific project ever performed by humanity was officially born.

In the agreement, Llewellyn Smith was careful to include the condition that any new financial support from countries not belonging to CERN should only be used to speed up the project and not to reduce the contribution from the member states. CERN sought help from outside its borders, and Japan, India, Russia, Canada and the USA answered the call. With their contributions, it would have been possible to manufacture and install all the dipole magnets together, and start the LHC at optimal conditions. But just as the single-phase option was becoming a reality, in 1996 a new crisis was triggered by the German decision to cut contributions to international scientific cooperation, in order to cope with its costs of reunification. A new process of negotiations between CERN and the member states started. The crisis was eventually resolved with the decision to allow CERN, for the first time in its history, to take loans. The single-stage construction of the LHC, with all magnets installed simultaneously, was eventually approved at the end of 1997.

In 2000 the LEP project was completed and dismantled to make space in the underground facilities for the LHC installations. However, in September 2001 the CERN management, led by the then director general Luciano Maiani, suddenly announced an increase in the cost estimate. The CERN Council was not ready to absorb the increase, and a programme of staff reduction and of redeploying resources towards the LHC became necessary. This programme, which was brought to completion by Robert Aymar, Maiani's successor, was successful in its goal: the construction of the LHC. But of course it came at a significant price: drastic cuts in the internal service and technical support, as well as reductions in research activities unrelated to the LHC. This was rather unfortunate, because a diversified scientific programme is essential for the intellectual vitality of a laboratory and for ensuring the development of new ideas and technologies. The LHC construction was eventually concluded with material costs amounting to about 3 billion euros, in excess of the original estimates by about 20 per cent. This is a remarkable success, considering the technological challenges and the cutting-edge research involved in the project. On 10 September 2008, the construction phase was officially completed when the proton beams made their first trip around the LHC ring. This event marked the start of the most exciting part of the project: the exploration of zeptospace.

The protons' voyage

> The only true voyage of discovery...consists not in seeing new land-scapes, but in having new eyes.
>
> Marcel Proust[6]

The collision of protons inside the LHC ring is only the final stage of a longer voyage. This voyage begins with the accumulation of protons obtained by stripping hydrogen atoms of their orbiting electrons. Then the energy of the protons is sequentially increased by a series of different accelerators: linac, proton synchrotron booster (PSB), proton synchrotron (PS) and super proton synchrotron (SPS). Protons are transferred from one accelerator to the other by fast-pulsating magnets (called kickers) that deflect their trajectories. Some of the accelerators used in this process are old glories of CERN. In their youth, they were marvels of their time, and they are the machines where celebrated experiments have been performed. The oldest is the PS, inaugurated in 1959, and even the SPS, where the *W* and *Z* particles were discovered, takes part in the operation of preparing protons for their fast rides in the LHC. All these old accelerators had to be upgraded and rejuvenated for the occasion. The process of preliminary acceleration, called the injection phase, is very delicate because the behaviour of the final beam depends critically on how protons have been initially treated (not unlike humans).

Once protons have completed the injection phase, they have reached an energy of 0.45 TeV and, at that point, they step into the main LHC tunnel. The tunnel – a legacy of LEP – has a length of 26.7 km and an internal diameter of 3.8 m, a very comfortable size to take a good, though rather monotonous, five-hour hike. Protons, on the other hand, complete a full turn in only 89 millionths of second. The tunnel does not follow a perfectly circular trajectory; there are eight arcs alternating with eight straight sections, each about 700 metres long, used for a variety of instrumentation.

The tunnel lies underground at an average depth of about 100 m. The ring is actually slightly tilted, with an inclination of 1.4 per cent, the depth and the tilt having been chosen essentially for geological reasons. The tunnel is excavated mostly in molasses, which is a compact rock composed of consolidated fluvial and marine deposits. At lower depths, there are moraine deposits of gravel, sand and loam which contain ground water and are inappropriate for underground construction. The slope of the tunnel, besides allowing excavations to remain within the molasses layer, also brought another benefit. At one side, the tunnel had to be linked to the SPS for injection, but at the other side it could be

---

[6] M. Proust, *In Search of Lost Time, Vol. V: The Captive*, 1923.

raised, since it lies at the foot of the Jura mountains. This helped to reduce the depth, and hence the cost, of the vertical shafts.

At the time of LEP the digging of the tunnel was delayed by a legal problem. Since in France (but not in Switzerland) the property of land-owners extends all the way to the centre of the earth, the excavations became possible only after the French authorities signed a "Déclaration d'Utilité Publique". The main reason why colliders are built underground is to have a natural shield against radiation. Moreover, it would be too expensive to buy all the surface land necessary to fit the huge LHC ring. Underground tunnels also reduce the impact on the landscape.

Two counter-rotating beams of protons are injected in the LHC ring in two separate pipes contained inside the *dipole magnets*. First-time visitors to CERN are often surprised to learn that the most expensive and technologically advanced part of the LHC is not what is responsible for increasing the proton energy, but rather for bending their trajectories. The role of the dipoles is indeed to bend the proton beam and maintain its circular orbit. There are 1232 dipoles inside the tunnel, all lined up with impeccable accuracy. Each of them is a 15-metre long tube painted an elegant sky blue. The length of the dipoles was actually determined by the mundane reason that it is the maximum allowed for transportation on European roads. Each dipole weighs 30 tonnes and costs about 700 000 euros. It is curious that, when expressed in euros

Figure 6.2 Dipole magnets installed inside the LHC tunnel.
Source: CERN.

Figure 6.3 Welding the interconnection between two LHC dipoles.
Source: CERN.

per kilogram, the price of the LHC dipoles – the most expensive part of the accelerator – is the same as Swiss chocolate. Were the LHC built of chocolate, it would cost about the same.

The LHC dipoles produce, in their interior, a uniform magnetic field that bends the two proton beams into their circular trajectories. The higher the energy carried by the protons, the harder it is to bend them and the more intense must be the magnetic field. Therefore, there are two options for reaching the highest possible energy in proton colliders: either make the diameter of the ring larger or increase the magnetic field. In the case of the LHC, the size of the ring was fixed by the LEP design and thus the maximum energy that protons can reach is determined by the intensity of the magnetic field inside the dipoles.

The LHC dipoles are designed to produce a magnetic field of 8.33 teslas, about 150 000 times stronger than the earth's magnetic field at Swiss latitudes. Enormous electric currents are needed to produce so intense a field. If such currents flowed in ordinary copper wires, they would rapidly dissipate more megawatts in heat than that which can be produced by a large number of power stations. So how can the LHC dipoles generate such an intense magnetic field?

The secret is an extraordinary physical phenomenon: *superconductivity*. Certain materials – called superconductors – have the odd property that, below a critical temperature, they conduct electric current with no resistance. Once a current starts flowing in a superconductor, it will keep flowing forever with no need of any battery. Superconductivity was discovered in 1911 by the Dutch physicist Kamerlingh Onnes (1853–1926, Nobel Prize 1913), and it is a phenomenon so unusual that it seems to defy the laws of electromagnetism.

In normal circumstances a current flowing through a conductor meets some resistance, dissipating energy in the form of heat. This phenomenon is exploited in all the electric heating appliances we routinely use at home. But this does not occur for superconductors, where currents can flow at no energy cost: no resistance and no dissipated heat. By running a superconducting hair dryer, you would get no heat no matter how intense is the current applied. Although superconducting hair dryers are probably not such a profitable invention, superconductivity has had many other interesting applications. It is nowadays employed to generate intense fields for magnetic resonance imaging, a diagnostic tool that visualizes the internal structure of the body. Superconducting cables can transport power with no energy dissipation and may become a reality in the future for energy storage, telecommunications, or electronic equipment.

Another stupefying property of superconductors is that they expel magnetic fields – a phenomenon called the Meissner effect. When a superconductor is placed in a magnetic field, some electric currents run across its surface. These currents produce a magnetic field which cancels the original one and screens the interior of the material from any externally applied magnetic field. If you place a piece of superconductor material on top of a magnet, you will see it levitating and floating in the air. The reason is that the magnetic field cannot propagate inside the superconductor because of the Meissner effect, and thus it effectively repels any superconducting material. The effect can be so spectacular that sometimes it looks more like witchcraft than science. But real science it is, and in 2003 a MAGLEV (magnetic levitation) train built in Japan reached a record speed of 581 km/h, travelling faster than the famous French TGV. Magnetic levitation trains can also run more quietly and more smoothly than ordinary wheeled vehicles and can potentially reach fantastic speeds, if operated in evacuated tunnels. Though not as fast as protons in the LHC, such trains could travel from Geneva to London in less than one hour.

The LHC has challenged the frontiers of superconducting technology, but it has also taken full advantage of the experience gained from HERA (the electron–proton collider built at the German laboratory DESY), from the Tevatron (the proton–antiproton Fermilab collider) and from the ill-fated SSC. The LHC dipoles contain coils of superconducting cables made of niobium–titanium. Each cable is made of stranded filaments 6 microns thick – about 10 times thinner than a human hair. It was a formidable industrial challenge to turn large bars of niobium–titanium into kilometres of these filaments, fulfilling the strict specifications and causing the minimum number of ruptures. The LHC is using more than a billion kilometres of these superconducting filaments – enough to wrap around Mars's orbit – for a total of 1200 tonnes of material.

When the LHC is in full operation, currents as high as 12 800 amperes flow through the superconducting cables. Just for comparison, the current flowing through the wire of an ordinary light bulb is less than 0.3 amperes. Thanks to such extremely high currents, the superconducting coils wrapped inside the dipoles can produce magnetic fields stronger than 8 teslas, able to steer the fast-moving protons within the racetrack of the LHC tunnel. The magnetic field is so intense that it exerts a violent force on the superconducting coils. Some of the force is supported by the geometrical configuration of the coils, just as in the case of a Roman arch. But a residual magnetic force tends to break the structure apart. This force is equivalent to a weight of 400 tonnes per metre – as if more than a thousand African elephants were sitting on top of each dipole. Specially constructed collars, made of 4-centimetre thick steel, sustain most of this tremendous force, while the rest is supported by the external structure of the dipoles.

There is however a small catch in using superconducting technologies. Superconducting materials lose their magic property above a critical temperature, and this temperature is terribly low. The LHC dipoles must be kept at a temperature of –271°C (just 1.9 degrees above absolute zero) or, in other words, colder than any of the emptiest places in outer space. Although this is another good reason not to market superconducting hair dryers, it has not stopped the construction of the LHC.

The daunting task is not just to reach these extreme temperatures, but especially to maintain, at –271°C, all 37 000 tonnes of material spread along 27 kilometres. Once again, an extraordinary physical phenomenon comes to the rescue: *superfluidity*. Under certain conditions of temperature and pressure, liquid helium becomes superfluid, completely losing its viscosity. It can flow freely, conducting heat 3000 times better than copper. This property allows physicists to keep the inner parts of the dipoles at extremely low temperatures, since superfluid helium is able to absorb any miniscule quantity of generated heat and to transport it efficiently outside the coils. The interior of the dipoles is bathed in liquid helium and the material chosen to insulate the coils is porous, so as to enable the helium to come in direct contact with the superconducting wires. At the same time, liquid helium is flowing through the dipole in a heat-exchanger pipe, which is used to maintain the system below the critical temperature for superconductivity.

The refrigeration (or *cryogenic*) system used to bring the temperature inside the dipoles to less than –271°C is the largest in the world and involves several stages of cooling. Refrigerator turbines and about 10 000 tonnes of liquid nitrogen are used, during the first stage, for cooling 130 tonnes of helium. This in turn circulates inside the dipoles, keeping the coils below the critical temperature for superconductivity. The process of cooling a sector of the LHC to the required temperature of −271°C takes nearly a month. During this time, metal parts undergo shrinkages and deformations that have been carefully taken into account during design and manufacture of the dipoles, since precise positioning during operation is essential. At the end of the cooling process, each dipole becomes several centimetres shorter, but the position of the coils must be accurate within a tenth of a millimetre. This watchmaker precision in the positioning of the thousands of kilometres of superconducting cable is necessary to obtain the required properties of the magnetic field. Indeed, the magnetic field inside the dipoles must not only be extremely intense but must also be precise and uniform in order to correctly guide the proton beam.

The LHC works close to the edge of the superconducting phase. This means that any small increase in temperature can bring the system to a state where superconductivity is lost. Every miniscule impurity present in the coils or even movements of the order of microns can produce a small amount of heat sufficient to raise the temperature of the superconducting material above its critical value. Then, like a broken spell, the miracle of superconductivity suddenly vanishes and the cable becomes resistive to the electric current. When this happens, it is said that the magnet has *quenched*.

As in a storm, when static electricity is discharged through lightning, the energy stored in the magnet is suddenly released after the quench. The LHC has a protection system against damage caused by the quench. As soon as a voltage exceeding 100 millivolts is detected between two ends of a dipole for a time longer than 10 milliseconds, the system is alerted. This is a warning that something must be wrong, because superconductors should have no resistance and thus no voltage between any two points should be registered. The presence of a voltage is the sign that the material is no longer superconducting and that a quench has started. In the case of quench, the priority is to quickly extract and dissipate the energy in a controlled way. Special heaters bring the entire dipole out of the superconducting phase, spreading the quench and distributing the energy release. At the same time, the current is immediately switched off. The process takes less than 200 milliseconds.

Inside the LHC dipoles, everything is at extreme conditions. In addition to the electric current, the magnetic field, and the temperature all reaching extraordinary values, there is another quantity brought to an extreme: the vacuum. A vacuum is required to provide the necessary thermal insulation for the magnets and the helium distribution line according to the same principle of the thermos bottle that keeps coffee hot but, especially, gas has to be evacuated very efficiently from the path of the proton beam. Molecules of residual gas inside the beam pipe are a threat because protons can collide with them, disrupting the beam stability. It is then necessary to pump out the air and reduce the pressure in the beam pipe to 10^{-13} atmospheres. To find an equally rarefied atmosphere, one should travel on a weather satellite orbiting the earth at an altitude of 1000 kilometres. The amount of space to be evacuated in the insulation vessels is staggering: about 9000 cubic metres, the volume of a theatre hall.

The dipoles are truly the pride of the LHC project. All their precious elements are contained inside the blue cylinders: the two pipes where the counter-rotating beams circulate; the superconducting coils held by steel collars; the helium heat-exchanger pipe; the vessels for the insulation vacuum necessary to maintain the low internal temperature. All these elements are held together with exceedingly accurate positioning, maintained even under the violent stress on materials caused by the intense magnetic field and the low temperature. The construction of the dipoles meant challenging the frontiers of many different technologies simultaneously. But the production of 1232 dipoles posed a significant industrial challenge as well. Although designed at CERN, they could not be manufactured inside the laboratory. This called for a close partnership with industry.

After a first phase of prototyping inside the laboratory, CERN selected three companies for the production process: one in France, one

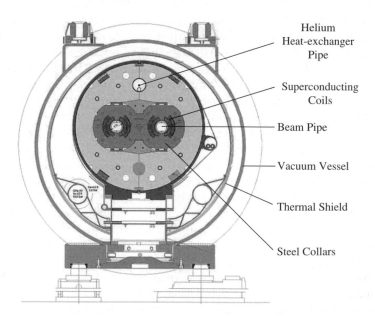

Figure 6.4 A schematic view of the cross section of an LHC dipole magnet.
Source: CERN.

in Germany, and one in Italy. Teams of CERN physicists, engineers, and technicians worked with the companies during a training period. Then, in 2000, an initial order for 30 dipoles was placed with each company. This helped the companies to gain experience, to improve production efficiency and to establish confidence. In this way the final order, issued in 2002, could be obtained at a much lower price, actually between a third and a quarter of the prototype cost, although after long negotiations.

One apparently simple, but industrially challenging, aspect was that the dipoles are not perfectly straight, but must be almost imperceptibly curved to follow the arc of the underground ring. In a total length of 15 m, they are bent by just 9 mm. Industries found a way to weld the dipoles under a large press capable of bending them. However, the accuracy of this automated procedure was too poor for the strict LHC requirements. It was then necessary, during the installation phase, to calibrate the central supports of the dipoles to achieve the precise curvature.

For the construction of the dipoles, CERN decided to take responsibility for procuring the main components to be assembled and even the raw materials, which were then delivered to the three companies. This helped CERN to maintain a close control over quality, uniformity, and cost. However, it entailed a considerable burden in organizing schedules, transport, storage, and logistics. CERN moved 120 000 tonnes of material around Europe, and on average 10 heavy trucks a day, for more than four years, were crossing Europe carrying dipole material. The close monitoring of the industrial operation and CERN's direct involvement made sure that each of the 1232 dipoles is virtually identical to the others and could be installed at any place of the ring without worrying about who had manufactured it or where the parts had come from.

Sharing the laboratory expertise with private industry was crucial to meeting the precise specifications for the various components. When considering the spin-offs of large projects in pure science, one should not forget the benefit to industry coming from the need for new manufacturing techniques. Many of the companies that worked for the LHC project are now using the new skills learned in the process. For instance, one company is producing superconducting material for medical magnetic resonance imaging and another has applied a special production process started for the LHC to manufacturing automobile parts.

After being delivered to CERN and carefully tested, each dipole was lowered through a vertical shaft into the tunnel and then transported to its proper location by a special vehicle. The vehicle was automatically guided by an optically detected line painted on the floor, as there were only a few centimetres of clearance between the wall of the tunnel and the LHC installations. To limit vibrations, the vehicle advanced at 2 km/h, slower than normal walking speed. This meant that the

Figure 6.5 Descent of an LHC dipole into the tunnel.
Source: CERN.

transportation of just one dipole to a location in the ring opposite to the shaft was a job lasting about seven hours. The project of the LHC dipoles was by no means a quick business. Lucio Rossi, the leader of the CERN group for magnets, cryostats and superconductors, once told me that dipoles followed the seven-year rule, first demonstrated by Marilyn Monroe in "The Seven Year Itch". The project took seven years of research and modelling (1988–1994), seven years of prototyping and industrialization (1995–2001), and seven years of construction and installation (2002–2008). The final dipole was lowered into the tunnel on 26 April 2007, with a banner saying "Magned olaf yr LHC". While everyone present at the ceremony applauded, only the Welsh LHC project leader Lyn Evans understood that it meant "Last magnet for the LHC".

Towards the final blast

> The infernal storm, eternal in its rage,
> sweeps and drives the spirits with its blast.
>
> Dante Alighieri[7]

When protons first enter the LHC ring, their energy is "only" 0.45 TeV. How are protons accelerated to their final energy of 7 TeV? Each proton beam encounters along one of the straight sections of the tunnel eight *radio-frequency (RF) cavities*, which look like shiny cylindrical tanks resembling massive water heaters. Inside the RF cavities, an electric field oscillates with a frequency of 400 MHz, the same frequency at which remote controls unlocking car doors operate. When protons enter the cavity, a pulse of the oscillating electric field gives them a gentle kick, and their energy increases by 485 billionths of TeV at each turn around the ring. This may seem very little, but protons complete 11 000 turns of the LHC ring every second and they receive a small kick at each turn. It is just like when you push a child sitting on a playground round-about. Just a very gentle push, though repeated at each turn, is sufficient to eventually make the roundabout spin so fast that the child will soon get dizzy. In the same way, protons acquire a very small amount of energy at each turn around the ring, while dipoles adjust their magnetic field to keep the beam within its trajectory. It takes about 20 minutes for the proton beam to reach the final energy of 7 TeV.

When first injected into the LHC ring at the energy of 0.45 TeV, protons already travel almost as fast as light – at a speed equal to 99.9998 per cent of the speed of light. But, according to special relativity, no particle can travel faster than light, so a large increase in energy amounts

---

[7] D. Alighieri, *The Divine Comedy, Inferno*, Canto V, 31–32, translated by M. Musa.

Figure 6.6 The radio-frequency cavities in the LHC tunnel.
Source: CERN.

to only a marginal gain in velocity. At the end of the acceleration process, protons have become 15 times more energetic, but their velocity has increased only by 0.0002 per cent. At that point they travel only 10 km/h slower than light. As the beam is accelerated, the frequency in the RF cavities is slightly modified in order to keep up with the small change in velocity of the proton bunches and to make sure that the pulses in the cavities do not to miss the right moment to push. During the full acceleration process the 400 MHz frequency is changed by less than 1 kHz.

The beam of protons is not uniform like a stream of running water, but rather the protons are grouped in bunches like water drops dripping from a leaky tap. Each bunch contains about 100 billion protons – the equivalent of 10^{-13} grams of matter. When the LHC is fully loaded, there are 2808 bunches rotating around the ring in each beam. As it first enters the tunnel, each bunch is about 10 cm long and 1 mm wide – like the lead of a pencil – and separated from the next bunch by a distance of about 10 m. The structure of the bunch changes with energy and becomes 7 cm long, once the energy of 7 TeV has been reached.

Protons too appear very different after they have been accelerated. When moving very fast, protons look dreadfully squeezed along the direction of motion, and yet their size in the orthogonal direction is not modified at all. Thus, the speedy protons at the LHC look like flat disks, like pancakes, whose thickness is about 7000 times less than their width. They have the proportion of pancakes less than a millimetre thick, like crêpes artfully cooked by the best French chef. What has happened to them?

The answer can be found in Einstein's special relativity. Odd as it may seem, if you measure the length of a body in motion with respect to you, you will find it contracted along the direction of motion. The faster it goes, the shorter it becomes. This effect is known as *Lorentz length contraction*, from the name of the Dutch physicist Hendrik Lorentz (1853–1928, Nobel Prize 1902). This phenomenon explains why, once protons have gained speed, they look squeezed. If a man could ride on top of a proton in the LHC, his measurement of the proton length would not change as speed increases. However, he would see the tunnel around him (and any physicist idly standing in the area) shrink along the direction of motion, as if viewed through a distorting mirror. Even more strangely, time would flow differently for our imaginary proton rider than for us. If he measured the time it took him to complete a turn of the LHC ring, his result would be some 7000 times less than what physicists measure in the laboratory. These are the strange things that happen in relativity.

The proton beam revolves around the 27 km tunnel in evacuated space. However, the vacuum cannot be perfect and indeed even at the pressure of only 10^{-13} atmospheres there are about 3 million molecules of residual gas for every cubic centimetre. Protons occasionally hit the gas molecules and are deflected out of their bunch. This can be a dangerous business if enough protons reach the coils of the superconducting magnets. There, misguided protons will deposit energy and possibly heat the material above the critical temperature, triggering a magnet quench.

It is therefore absolutely necessary to intercept protons that have gone astray. But catching in flight a 7 TeV proton is no trifle. All along the LHC beam there is a system of collimators, made of carbon material that can survive impact with energetic protons. Collimators are like the teeth of an alligator's jaw, slightly open to let the beam go through, but ready to catch any straying proton. Their purpose is to "clean" the beam by removing particles that are not well arranged inside the bunch. The system of collimators is movable and the jaws can open or close upon the beam. In normal running conditions, the opening between collimators – the aperture through which the beam travels – is about 3 mm wide.

The intensity and stability of the proton beam deteriorates with time, because of collisions, particle losses, and the motion of protons within the bunch. This happens, on average, after about 10 hours when the beam has circulated around the LHC ring some hundred million times and covered a distance equal to crossing the solar system from one side

Figure 6.7 Digging the tunnel that will be used to host the dump block.
Source: CERN.

to the other. When the beam shows signs of old age – or in case of an emergency – a kicker magnet deflects the protons, directing them into the *dump block*, which is a cylinder of graphite composite 8 m long and 1 m in diameter, encased in concrete. The dump block is the only element of the LHC able to withstand an impact with the full high-energy beam.

In order to guide the beam around the tunnel, the LHC contains thousands of magnets other than the dipoles (such as the *quadrupole magnets*). These magnets are needed to correct instabilities, optimize the trajectory and focus the beam. Essentially they operate in the same way that optical lenses focus light rays. But, in contrast with ordinary optical lenses, magnetic focusing acts only in one direction: if the beam is focused on the horizontal plane, it gets defocused along the vertical plane, and vice versa. That is why a succession of focusing and defocusing magnets are necessary to obtain the desired squeezing of the beam.

During most of its journey inside the tunnel, the proton beam has a diameter of about a millimetre. But at four points around the ring, the two counter-rotating beams cross each other and protons are brought into head-on collisions. As they approach these points, the beams are squeezed through magnetic focusing. At this stage, the beam has a diameter of only 16 microns – thinner than a human hair. This last squeezing is crucial to make the beam more intense and increase the chance of collision. At this point protons are ready to spring into action for their great final crash.

The two beams are not brought together fully head-on, but at a small angle, in order to avoid unwanted simultaneous collisions between different bunches, which would deteriorate both the properties of the beams and the data taken by the experiments. At the moment of collision, the two beams cross each other at an angle of 280 microradians. This is a very small angle; it is equal to the angle subtended by a one-metre tall object viewed from a distance of 3.6 kilometres.

The fatal moment has finally arrived. A hundred billion protons forming a front only 16 microns across come from one side at dazzling speed and ram against an equally numerous battalion of protons dashing forward to meet them. And what happens next? Actually, most of the protons miss each other and go through completely unscathed!

This, perhaps disappointing, outcome is just the consequence of the smallness of protons. Each of them has a size of about a millionth of a nanometre and the probability of knocking against another approaching proton is very small, even in spite of having so many of them. Nevertheless, a few crashes among protons occur, creating the kind of collisions that physicists are eager to scrutinize. In fact, the intensity of the proton beam at the LHC is just right to produce an adequate number of collisions. If a hundred billion protons simultaneously crashed, each creating hundreds of new particles in the debris of the collision, the result would be an inextricable mess and any measuring instrument would immediately overflow with an excess of data. Physicists want enough collisions to observe the rare events they are interested in, but not too many to be swamped by tangled information.

On energy, safety and the unforeseen

> Energy is eternal delight.
> William Blake[8]

How much energy is involved in a collision between two protons at the LHC? If you expect some colossal number, be ready for a disappointment. The energy carried by two colliding 7 TeV protons is the same as the kinetic energy of two mosquitoes flying into each other. The energy released in the proton collision is equal to the energy released by the sound of a light knock on a door. So what is the big deal about the LHC? The special property of the LHC is to concentrate this energy in a very small space. Instead of being carried by a flying mosquito or propagated

---

[8] W. Blake, *The Marriage of Heaven and Hell*, 1790.

by the sound wave in a room, all that energy is squeezed into a slice of zeptospace. This is what makes the LHC so powerful.

Although the energy released in a proton collision is small by normal standards, the total energy stored in the underground tunnel when the LHC is running is considerable. After all, the LHC absorbs about 120 megawatts of power (operating the accelerator and all detectors included). This is comparable to the household electricity consumption of a town the size of Geneva. The energy carried by a single proton is rather insignificant, but there are about one hundred billion protons in each bunch, and 2808 bunches in each beam circulating at the LHC. Thus, the total energy of the proton beam is 0.36 gigajoules, which is equivalent to the kinetic energy of a 400-tonne TGV train travelling at 150 km/h. The beam has to be steered very carefully: losing the beam is like letting a TGV train run wild. If misdirected, it has enough destructive power to melt a half-tonne block of copper. That is why only the specially constructed dump block can sustain the direct impact of the beam.

An even larger amount of energy is stored in the dipole magnets: a total of about 10 gigajoules – equivalent to 2.4 tonnes of TNT explosive. Because of the extraordinary amount of energy stored in the dipoles, a protection system in case of magnet quench is an essential element of safety. But, in case of emergency, what really matters is not the total amount of energy, but the way in which that energy is released. For instance, instead of comparing 10 gigajoules to 2.4 tonnes of TNT, we could have said that the energy stored in the dipoles is equivalent to the calories contained in 460 kg of chocolate. The comparison sounds a lot less destructive. By gathering enough hungry children, we could get rid of all that chocolate, releasing the energy in a (relatively) safe way. The LHC quench protection system works according to the same logic: it dissipates energy in a diffused and controlled manner.

An important issue in a project the size of the LHC is to predict the unforeseen and take all possible precautionary measures to avoid any kind of dangerous accident and to ensure absolute safety during operation. CERN has deployed great resources on this issue. However, although safety has always been guaranteed, it would be unrealistic to expect everything to go smoothly when dealing with prototype technology. A series of unexpected delays and accidents are unfortunately unavoidable in a project with the complexity of the LHC.

In the summer of 2004 it was discovered that the cryogenic distribution line – that is the system feeding liquid helium into the magnets to keep them cold – was defective. The components, supplied by an external company, were not up to the required specifications: some of them were faulty and the welding was of poor quality. Production had to be immediately stopped. CERN worked in close collaboration with the company to fully redefine the manufacturing process and the quality-control tests during production. All defective components already installed in the

LHC had to be replaced and the operations amounted to a delay in the schedule of about one year.

On 27 March 2007, during a high-pressure test in the tunnel of the LHC, one of the "inner triplets" – systems of three focusing magnets devised to squeeze the beam of protons before the collision point – burst, damaging the nearby electrical connections. The inner triplets, built outside CERN, were not designed to withstand the asymmetric force that was applied during the test and that could be exerted in case of some accident. The support structures had to be redesigned, but it was possible to install them without removing the undamaged inner triplets from the tunnel, an operation that would have cost a much longer delay in the schedule.

The most recent accident occurred on 19 September 2008. The accident was reported in the media, especially because it happened just after the successful event of 10 September, when the proton beams made their first triumphant turns around the LHC tunnel. At that time, almost all of the dipoles of the LHC had already been tested up to an electric current flowing through the superconducting coils of 9300 amperes, which allows acceleration of the proton beam up to energies of 5.5 TeV. However, in one sector the current had been tested only up to 7000 amperes. On the fateful morning of 19 September, it was decided to complete the test of that sector by ramping up the electric current.

At 11.18 a.m. the screens of the monitors in the LHC Control Centre became red with alarm warnings. Mechanical damage in one of the dipoles had caused a leak of liquid helium. Helium initially at $-271°C$ immediately vaporized as soon as it escaped from its enclosure. Two tonnes of helium were released in less than two minutes, propagating at an initial speed of about 70 km/h; then the leak continued less violently for a total loss of six tonnes of helium. At the moment of the accident, the sensors for oxygen deficiency and the fire alarms along the tunnel went off in rapid succession, and this is actually how the speed of the helium front could be measured. There was no fire but the smoke detectors are sensitive to optical transparency and the burst of helium had lifted a cloud of dust. Needless to say, nobody was present inside the tunnel because access is strictly forbidden during such operations.

The shock wave produced by the release of helium in the dipole vacuum vessel displaced 39 of the 30-tonne dipoles and many other magnets from their accurately aligned positions, crashing some of their connections. The rupture of the helium enclosure was most likely caused by an electrical fault due to a defective splice connecting two dipoles, although it is impossible to assess the chain of events with absolute certainty, because the implicated connection was completely vaporized during the accident. CERN has responded to the incident with earnest professionalism and hard work, in spite of the bitter disappointment. Help has also come from Fermilab, which sent a team of experts to speed up the repair work. Inside and between the dipoles there are about 24 000 splices similar to the one that presumably caused the accident. Physicists and engineers have quickly developed clever schemes to look for any other possible defective splices. But the work of repair and consolidation, necessary to prevent reoccurrence of similar accidents, has delayed the schedule by at least one year, as the LHC is planned to restart operations at the end of 2009.

7
Telescopes Aimed at Zeptospace

We are all in the gutter, but some of us are looking at the stars.

Oscar Wilde[1]

Had we no telescopes, we could not observe very distant supernovae explosions. In the same way, we need special instruments to observe and study the miniscule explosions of particles originating from the collisions between protons at the LHC. In physics, these particle explosions are called *events* and the instruments to observe them are called *detectors*. Detectors are built to record the echoes from zeptospace. They register the event by reconstructing the tracks of all particles moving away from the beam and by measuring their properties, such as electric charge, energy, and momentum.

At the LHC, detectors are placed in underground caverns located at the four points where the two counter-rotating proton beams intersect. The two main detectors are called ATLAS (A Toroidal Lhc ApparatuS – a laboured but catchy acronym) and CMS (Compact Muon Solenoid – a precise but dull acronym), and they cost about €300 million each. They are located in two caverns at opposite points of the LHC ring, ATLAS in Switzerland and CMS in France. The pre-existing caverns used for the LEP experiments were too small to fit the gigantic instruments of ATLAS and CMS, so new ones had to be excavated. Visiting the ATLAS cavern – the larger of the two – in 2003, soon after the civil engineering work ended, was quite an impressive experience: an empty space the size of a cathedral, 100 m underground, connected to the surface by breathtaking vertical shafts. Now the cavern is packed full with the massive detector.

Excavations led to an interesting surprise. In the CMS area, remains of a Gallo-Roman villa from the fourth century AD were found. CMS, which is supposed to explore the physics of the early universe, was starting on the right foot with a first discovery back in time, although

---

[1] O. Wilde, *Lady Windermere's Fan*, 1892.

not as far back as the Big Bang. Excavations brought to light coins minted in *Ostium* (Ostia), *Lugdunum* (Lyon), and *Londinium* (London). British colleagues could not resist commenting that when they arrive in the area carrying only pounds, but no Swiss francs or euros, there is no way for them to get any food in supermarkets, neither in Switzerland nor in France. Apparently their ancestors had no such difficulties: globalization has really made giant steps.

Excavations for CMS were slowed down when the drilling hit the water table. The cavern started to flood and pumping out water would have been slow and inefficient. At that point, engineers installed a system of vertical pipes buried in the ground with a double cooling circuit filled with ammonia and salt water at −23°C. At a later stage, liquid nitrogen was circulated in the pipe system in order to completely freeze the water table. Drilling through ice is no more difficult than drilling through rock, and so in this way the work could be completed. Another problem presented by the excavations was that the rock around the CMS cavern was not sufficiently hard, and so supporting structures had to be built. These operations, though successful, amounted to a considerable delay of the civil engineering work.

After the completion of the ATLAS cavern, with the removal of 300 000 tonnes of rock, the floor of the cavern began a slight upward movement – almost 1 mm per year. This movement had to be constantly monitored by a very sensitive metrology system, to ensure the precise alignment of the detector components. The metrology instruments in the cavern were so sensitive that they were able to detect the tsunami of December 2004, recording the Macquarie Island, and the Sumatra earthquakes, as well as the subsequent tidal wave.

ATLAS and CMS are called "general-purpose detectors" because they are ready for any kind of result from proton collisions. Indeed, they record the full information about the collision event by identifying all produced particles and reconstructing their trajectories (except for a small cone along the beam direction). In practice, they take a complete snapshot of every event. Since identification of different particles requires different techniques, ATLAS and CMS are actually an assembly of many different instruments, each one with a specific purpose. All of these instruments are put together in a single gigantic structure. ATLAS is 46 m long and 26 m high or, in other words, bigger than Solomon's Temple (at least according to rabbinic tradition).

Detectors for the LHC have to satisfy strict requirements that posed difficult technological challenges. First of all, the response of electronic components has to be very fast, because the time lag between collisions of two proton bunches is only 25 nanoseconds. Secondly, all equipment has to be resistant to high doses of radiation, since it is constantly exposed to large fluxes of energetic particles, especially in the inner parts of the detectors, close to the collision point. Finally, the instruments have to be

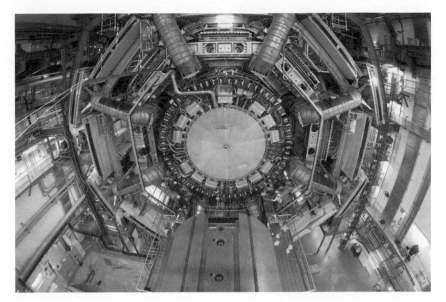

Figure 7.1 The ATLAS detector before the installation of the end-cap.
Source: CERN / ATLAS Collaboration.

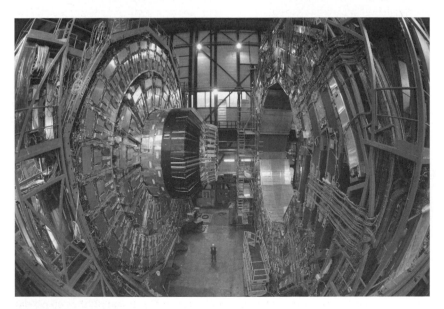

Figure 7.2 The CMS detector in the surface laboratory.
Source: CERN / CMS Collaboration.

well tested and reliable, since no repair or maintenance is allowed during operation, when access to the underground areas is forbidden. In addition, even during shut-down periods, replacing any equipment inside the detectors is extremely difficult and time-consuming. For this reason, everything is designed to survive for at least ten years without human intervention. In this respect, experiments at the LHC are not unlike missions in space. Considering all these requirements, it is not surprising that detector designs required many years of research and development, with a long process of selecting and producing special materials, and of instituting severe quality controls.

In order to interpret the results from proton collisions, physicists need to have information of all the particles produced in the event. Therefore, the detectors have to cover any direction around the collision point, save two holes where the beam goes through. To express this requirement, physicists say that detectors have to be "hermetic". One may think that a sphere would be the optimal shape, but in reality the LHC detectors look like gigantic cylinders, with the axis along the beam direction and one *end-cap* on each side to make them as hermetic as

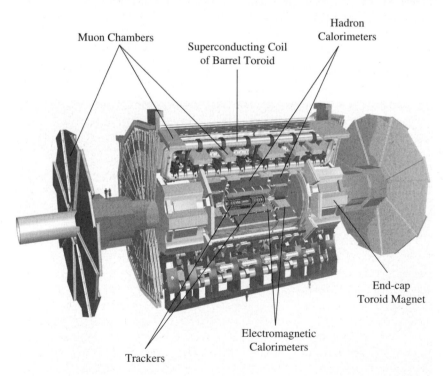

Figure 7.3 A schematic view of the ATLAS detector. The human figures above the left portion of the detector give an indication of the scale of the drawing.

Source: CERN / ATLAS Collaboration.

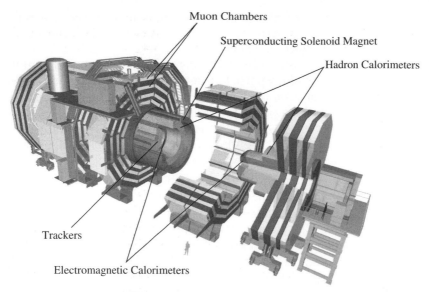

Muon Chambers

Superconducting Solenoid Magnet

Hadron Calorimeters

Trackers

Electromagnetic Calorimeters

Figure 7.4 A schematic view of the CMS detector. The human figure in the foreground gives an indication of the scale of the drawing.

Source: CERN / CMS Collaboration.

possible. This geometrical shape is chosen for simplicity of design and for ensuring a uniform magnetic field inside the detector.

The ATLAS and CMS detectors are extremely complicated instruments and each of them employs different techniques to perform the various tasks. However, there are four main structures that are common to both detectors and constitute their backbones. We will examine them, following ideally the path of particles produced in the collision between protons, starting from the collision point in the heart of the detector and moving outwards.

1. Trackers. The trackers are composed of several different instruments contained in the innermost part of the detectors and are the first set of equipment met by the particles bursting out from the proton collisions. They are the most elaborate part of the detector with an incredible number of sensors, thousands of connections per square centimetre and millions of electronic channels. The trackers are made mostly of thin layers of silicon connected to layers of electronics. When a charged particle goes through one of the silicon layers, it liberates electrons, which are detected by the electronics as an electric current and then converted into a digital signal. This gives precise information on the position at which the charged particle has crossed a silicon layer. By joining together the information from different layers, the particle trajectory is reconstructed. The layers in the trackers have to be very thin in order not to divert particles from their natural paths. This detection method is based on electromagnetic interactions, and it is sensitive only to electrically charged particles. Neutral particles – like neutrons or photons – are invisible to trackers.

The trajectory measured by trackers gives some preliminary information on the nature of the particle, but this is insufficient to determine the particle's identity completely. Trackers act like a host who greets some foreign guests at a party. In the entrance hall, the host first asks some polite questions about their nationalities or their professions, receiving only some quick preliminary information, before the guests move into the next room to allow space for new visitors.

2. Electromagnetic calorimeters. The electromagnetic calorimeters are the next step in the journey made by particles produced in the collisions. Here electrons and photons come to a stop, releasing their energy into the material. The calorimeter promptly measures the amount of energy deposited by the particles and registers the information. At this stage electrons can be easily distinguished from photons. Indeed, although both kinds of particles are stopped inside the electromagnetic calorimeter, photons leave no trace in the trackers, while electrons do. Thus, at this stage, electrons and photons are fully identified. If particles inside the detector were guests in a carnival masquerade ball, masks from the faces of photons and electrons would have fallen by now, exposing their true identities.

Figure 7.5 The CMS inner detector during assembly.
Source: CERN / CMS Collaboration.

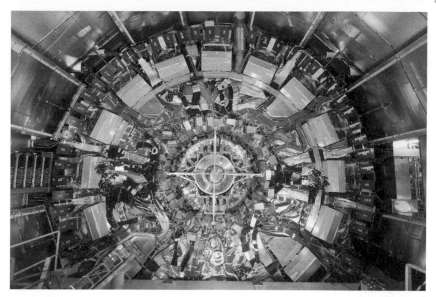

Figure 7.6 The ATLAS inner detector end-cap.
Source: CERN / ATLAS Collaboration.

ATLAS and CMS use different techniques in their electromagnetic calorimeters. ATLAS uses layers of lead arranged in the shape of an accordion filled with liquid argon at −186°C. When electrons or photons produced in the proton collision hit the metal layers, they create showers of lower energy particles. These particles liberate electrons from the atoms of liquid argon. The total electric charge freed in this process gives information on the energy of the initial particle. Argon is a noble gas well suited for particle detectors because it does not react chemically with other elements. Krypton, another noble gas, was also considered for the ATLAS calorimeter, as it leads to better energy resolution. Eventually krypton was discarded, not so much because kryptonite can be lethal to Superman, but rather because it is more expensive and the purification process causes some difficulties.

The electromagnetic calorimeter of CMS is instead based on a special material: scintillating crystals of lead tungstate. These crystals are stylish-looking small bricks, perfectly transparent like glass. However, by lifting one of these bricks, one immediately realizes that they are not made of common glass: they weigh more than pure iron. The crystals have the property of being very resistant to radiation and of allowing extremely precise determinations of the energy of photons and electrons. This property is very important for making accurate measurements of great relevance to the hunt for the Higgs boson.

Research on this special material had been conducted at CERN, but the production of the 78 000 crystal bricks contained in the CMS detector was carried out in two chemical factories: one in a nearly disused Russian complex, previously supplying the Soviet Army, and the other one in China. The procedure for manufacturing the crystals starts with melting salts containing lead and tungsten inside platinum ovens. Part of the platinum used for the ovens was borrowed from Russian and Swiss banks and then returned at the end of the production, after a purification process. A microscopic crystal attached to a rod is then inserted inside the oven and moved extremely slowly through the liquid lead and tungsten. This catalyzes the crystallization process, making the crystal grow. In the Russian factory, the artificial growth of each crystal lasted about two days. The Chinese factory followed a different procedure, in which the growth took about twenty days, but many crystals could be produced simultaneously. After a crystal had reached a length of about 20 cm, it was cut and polished using discs covered with diamonds, which is the hardest naturally occurring material.

During the nearly ten-year period in which the crystals were produced, CERN had direct experience of the transition in the Russian economy. The factory, at the beginning heavily subsidized by the state, underwent the move towards a free-market economy. Energy costs grew enormously, generating moments of crisis and contract renegotiations with the company. Eventually, the final orders to the Russian manufacturer had to be made in roubles, and not in dollars, because the company considered the Russian currency to be more stable and stronger than the US dollar.

3. Hadron calorimeters. Hadrons – particles made up of quarks and gluons, like protons, neutrons, and pions – mostly penetrate through the electromagnetic calorimeter. Then they reach the next stage of the detector, the hadron calorimeter, where they finally come to a halt. Here hadrons are stopped by metal absorbers, and their energies are detected by tiles of plastic scintillators, made of a material which then radiates light when exposed to charged particles. From the intensity of light, it is possible to measure the energy of the hadron.

A peculiarity of hadrons is that, when produced in proton collisions, many of them travel tightly together in a stream of particles, like drops of water in the jet from a hose. This special behaviour of hadrons is a direct consequence of their strong interactions.

In the violent burst following a collision at the LHC, single quarks and gluons are ejected from the interior of the protons. But QCD – the theory of the strong force – does not allow for the propagation of free quarks and gluons. What happens then? The metaphor introduced in Chapter 4 to explain how QCD confines quarks inside the proton could again be useful to illustrate the situation. When a quark is ejected in a collision, the rubber band that binds it to the other proton's constituents gets stretched. The more the rubber band is stretched, the more energy is stored in it. When this energy exceeds the mass of a hadron, energy can materialize in the form of a particle, according to Einstein's equation $E = mc^2$. At that point, the rubber band is stretched so much that it snaps and new hadrons are created. So, when two protons collide at the LHC, many of the elastic bands between quarks snap into pieces, leading to a stream of hadrons flowing in the direction of the original ejected quark. Collisions, no matter how energetic, cannot liberate individual quarks or gluons. QCD has condemned them to a life sentence inside the prisons of hadrons.

This process in which quarks and gluons form new hadrons is very complicated. It involves a quantum-mechanical effect in which quark–antiquark pairs materialize out of nowhere and recombine with the original quarks and gluons to create hadrons. If you have not understood what is going on, don't feel discouraged; theoretical physicists too are unable to give a full account of these processes. The problem is that, because of the peculiarity that the strong force becomes stronger as distance grows, the equations of QCD can be approximately solved in the limit in which quarks are very close together, but become too complicated when a distance equal to the proton size separates the quarks. So far nobody has ever been able to solve these equations. The confinement of quarks inside the proton has not been mathematically proved, but only reproduced through numerical simulations. Incidentally, exactly solving the equations of QCD would be an easy way to make some cash. The Clay Mathematics Institute of Cambridge, Massachusetts, has included this problem in its list of the Seven Millennium Problems and has offered a million dollars to whoever can solve it. The prize is still unclaimed.

However, in the exploration of the properties of zeptospace, we are more interested in the interactions of quark and gluons at very small distances, rather than knowing the behaviour of each individual hadron. Therefore, in the analysis of LHC data, all information about hadrons streaming closely together is combined into a single quantity, the *jet*. For physicists, a jet is not a powerful aeroplane, but a spray of hadronic particles, flying like a dense flock of birds. A jet is the signal for the production of either a quark or a gluon, but unfortunately it is not easy to distinguish between the two. Ongoing research is studying how to extract this information from the features of the particles composing the jet.

4. Muon chambers. Like those guests that keep on dancing and do not leave the party even when most of the others have gone home, muons continue their paths zipping through the hadron calorimeter. Muons are the most penetrating particles detected by ATLAS and CMS. Only neutrinos are more penetrating, so penetrating that they are totally invisible to detectors and cross their full volume without leaving a single trace.

Precise determinations of muon trajectories are made in the muon chambers, which are situated in the outer parts of the detectors. ATLAS and CMS use several different techniques to detect muons, but most muon chambers are made of small tubes filled with gas. As the muon passes through, it leaves a trail of electrically charged particles. These particles drift either towards a filament located at the centre of the tube, or towards the sides of the tube. From the time it takes for the charges to drift, it is possible to determine with great accuracy the position of the muon, as it passed through. The muon chambers in ATLAS cover an area of more than three soccer fields, but the precision with which they determine the muon trajectories is of hundredths of a millimetre.

Both the ATLAS and CMS detectors are built around powerful magnets that create strong magnetic fields in their interiors. These magnetic fields are needed to bend the trajectories of the charged particles produced in the collisions. From the way a particle bends in the magnetic field, it is possible to extract very important information. For instance, if the electric charge of a particle is positive it bends one way, if it is negative it bends the other way. Also, faster particles bend less than slower particles, so particle momentum can be measured from the curvature of the trajectory. The magnetic fields inside the detectors have to be extremely intense because energetic particles are hard to bend, and

Figure 7.7 The eight superconducting coils of the ATLAS barrel toroid before installation of the calorimeters and the inner detector.

Source: CERN / ATLAS Collaboration.

also because the stronger the magnetic field, the more accurate is the measurement of the particle momentum. Moreover, the magnets curl up the trajectories of low-energy particles, thereby acting like a filter that retains only the energetic particles, which are generally more interesting for data analysis.

ATLAS and CMS have chosen different schemes for generating the magnetic field inside their detectors. In ATLAS, the magnetic field is powered by a central solenoid and a colossal system with a *barrel toroid* and two *end-cap toroids* at the sides. The barrel toroid magnet consists of eight gigantic *superconducting coils* – shaped like oblong doughnuts – placed radially around the beam line. Being so large and visible, the toroid magnet system has become a trademark of ATLAS, and it is responsible for the "T" (toroidal) in the acronym. A sectional view of ATLAS looks like a gargantuan orange sliced in half, with the super-conducting coils delineating the segments.

The descent of the 25 m long and 5 m wide coils into the cavern, a journey of 100 m down the vertical shaft, was an impressive show to watch. The coils, suspended on cables, were lowered, then carefully rotated and positioned inside the detector. These giants were manoeu-vred around the concrete walls of the shafts and the cavern with only a few centimetres of clearance. If you have difficulties parking inside your garage without scratching your car, you would have been even more impressed by these operations.

The CMS magnet is a large *superconducting solenoid* inserted in a massive iron structure that contains the magnetic field. Not only is it responsible for the "S" (solenoid) in the CMS acronym, but also for producing an extremely powerful magnetic field in the interior of the detector, with an intensity of 4 teslas. This system allows the detector to be smaller than ATLAS, explaining the origin of the "C" (compact) in the CMS acronym. Nevertheless, the word "compact" isn't what first comes to mind when you face the CMS detector. It weighs 14 000 tonnes – like a fleet of about eighty Boeing 747-400s – and it contains more iron than one and a half Eiffel Towers. But it is indeed much more compact than either the aeroplane fleet or the Eiffel Tower, having a diameter of 15 m and a length of 21 m.

The powerful CMS superconducting solenoid magnet operates at a temperature of −268°C. The energy stored in the magnet is capable of melting 18 tonnes of gold. When it was first tested, it wiped out the hard disk of a laptop and erased a couple of credit cards of some imprudent physicists who went too close to it. It also blocked the elevator connecting the CMS cavern to the surface, because the magnetic field interfered with the electronic control system.

A visit to ATLAS or CMS in the underground caverns is an awe-inspiring experience. The most impressive aspect is of course the sheer size of the detectors, especially after considering that these enormous

Figure 7.8 Descent of the first superconducting coil inside the ATLAS cavern on 26 October 2004.

Source: CERN / ATLAS Collaboration.

Figure 7.9 Insertion of the trackers inside the CMS detector.
Source: CERN / CMS Collaboration.

apparatuses contain equipment aligned with micron accuracy and synchronized with nanosecond precision. Just glimpsing the indescribable entanglement of cables inside ATLAS or CMS gives an immediate visual impression of the utter complexity of these instruments. Thousands of kilometres of cables and optical fibres power the various parts of the detectors and extract digital and analogue information from the various instruments and from the myriad sensors constantly monitoring the equipment. This intricate system of cables resembles the circulatory system of some behemoth organism. Standing in front of these giants – with their complexity of microtechnology mixed with the enormity of colossal proportions – gives a sense of awe and amazement that is hard to describe. Even people with no special interest in science are moved to wonder at the grandeur of purpose and magnitude of scale when confronted with these powerful detectors. For me, the feeling surpasses that which one experiences by contemplating the pyramids of Egypt or any of the most spectacular monuments built by ancient civilizations. These detectors, like cathedrals of the 21st century, are the ultimate masterpieces of human ingenuity and the desire for knowledge.

Logistics and transportation

Music is a means of rapid transportation.

John Cage[2]

ATLAS was constructed with the "ship in the bottle" technique, in which each component was separately lowered and then assembled inside the cavern. On the other hand, CMS was almost entirely built and tested in the laboratory on the surface. This strategy, rather unusual in particle physics experiments, was decided upon to allow the assembly of the detector to start well before the civil-engineering work in the cavern had been completed. It also proved to have several other advantages. Maintenance and installation was much easier and, because of the available space, it was possible to work on different elements in parallel. The CMS detector was built in only 15 large slices that were lowered separately between November 2006 and January 2008. On 28 February 2007 the heaviest piece, weighing 1920 tonnes – as much as a herd of 400 African elephants – was lowered into the cavern. That piece, measuring 17 × 16 × 13 m, was lowered at an average speed of 10 metres per hour – literally slower than a snail – by a special gantry crane, built specifically for the purpose. Given that there were only 10 cm of leeway on each side between the CMS element and the walls of the shaft, a special system was monitoring and controlling any small sway of the cables down the 100 m descent. The hydraulic jacks and the crane employed in this delicate operation are going to be used again for a somewhat different purpose. They will lift the roof of a soccer stadium in Durban, South Africa, for the 2010 World Cup.

Most of the components of ATLAS and CMS were not built at CERN, but in laboratories and industries in every corner of the globe. Transportation of these colossal pieces of equipment was sometimes an epic adventure. The story of the two end-cap toroid magnets of ATLAS is a good example. Each toroid weighs 240 tonnes and has a diameter of 12 m. Its main component is an 80-tonne vacuum vessel, built by a Dutch company under supervision of physicists from the NIKHEF Laboratory, near Amsterdam, and from the Rutherford Appleton Laboratory near Oxford. Split in two halves, it was shipped down the River Rhine from the Netherlands to Strasbourg and then transported by a special truck convoy escorted by the police, travelling at 10–15 km/h. Because of the extraordinary size of the transport, the itinerary was carefully chosen to avoid tunnels, underpasses, or any such obstructions. It turned out to be necessary, however, to dismantle the high-voltage lines to cross a railway

---

[2] J. Cage, *A Year from Monday: New Lectures and Writings*, Wesleyan University Press, Middletown 1967.

Figure 7.10 Lowering of the heaviest piece of the CMS detector on 28 February 2007.

Source: CERN / CMS Collaboration.

track at one point. The journey from Strasbourg to Geneva took four days, but it was successful.

The surprise came with the shipment of the second end-cap toroid. Once the convoy reached the Jura Mountains, it found that the road was passing under a bridge that did not exist at the time of the first trip. The bridge had just been constructed to connect some hotels to the ski slopes. Cranes had to be sent from CERN to lift the end-cap magnet and swing it over the bridge while the truck passed under it, waiting to be reloaded. Around the time that this transport was completed, a radiation heat shield came to CERN from Israel, as well as thermal insulation material from Austria, and conductors from Germany, Italy, and Switzerland. The equipment was then assembled in the building with the largest door in the whole laboratory. The whole end-cap toroid magnet made its final trip to the entrance of the ATLAS shaft on a special trailer with 128 wheels, and then it was lowered into the cavern.

Over a glass of wine at the CERN cafeteria, physicists from ATLAS and CMS tell many curious stories about transportation of detector equipment. A foreign truck driver who had to make a delivery to "le Cern" (to be read with a French accent) drove instead to the city of Lucerne. A Swiss commuter, distracted by the impressive view of the transportation of one colossal LHC component, bumped into the car in front of him, creating the first LHC-related collision ever recorded.

Figure 7.11 Transport of the ATLAS end-cap toroid magnet, weighing 240 tonnes, on a special trailer with 128 wheels, on 6 February 2007.
Source: CERN / ATLAS Collaboration.

The ATLAS pixel detector, a precious electronic instrument with 80 million channels, travelled from Berkeley to CERN in a first-class airline seat. Secured in a high-impact plastic box, it was too big to fit on an economy airline seat, but it fitted just right in first class. It isn't recorded who drank the complimentary champagne that came with the first-class ticket.

Dealing with data

> If your experiment needs statistics, you ought to have done a better experiment.
>
> Ernest Rutherford[3]

At a Formula One grand prix, two cars that look quite far apart on the race-track – say 100 metres – can actually be separated by a rather short time interval of only one second. Something similar happens to the proton bunches circulating in the LHC ring. Bunches are separated by about 7 metres, and this may appear as a good deal of distance, when

[3] E. Rutherford, as quoted in N.T.J. Bailey, *The Mathematical Approach to Biology and Medicine*, Wiley, New York 1967.

compared to the size of a proton. However, the lapse between the arrival of two subsequent bunches at the collision point is only 25 nanoseconds. Protons go even faster than a Formula One Ferrari.

On average about 20 to 40 protons crash at every bunch crossing. This means that the rate of collisions (or of "events") at the LHC is utterly astronomical: around one billion events every second. At each event, hundreds of particle tracks are measured by the detector and converted into digital information. The total amount of data produced at the LHC is about a million gigabytes per second. This is sufficient to saturate every hard disk on the planet in about a day. But the LHC is not going to run for only a day, rather for many years. Why do physicists need to collect so many events of proton collisions at the LHC? How can they store this monstrously large amount of data?

The first reason for collecting data on a huge number of LHC events is that the vast majority of them are not very useful for the exploration of zeptospace. In most cases, when two protons collide, the quarks and gluons inside one proton do not come into direct contact with the quarks and gluons of the other. Quarks and gluons are actually much smaller than the size of the proton itself. We can picture them as microscopic seeds inside a large tomato. When two very ripe tomatoes smash against each other, pieces of pulp splatter all over the place, but the seeds remain in the pulp without having a direct hit. In jargon, these kinds of collisions are called *soft events*. In soft events, the energy of the collision is distributed over the proton and not concentrated in a much smaller region. Like collisions between two tomatoes, soft events do not probe the small distance scales of zeptospace.

Occasionally, quarks or gluons have direct head-on hits. These kinds of collisions are referred to as *hard events*. In a hard event, a large fraction of the available energy is concentrated in a very small point, and thus it can be fully converted into the creation of some unknown heavy particle. Physicists exploring small distances are interested in hard events.

Unfortunately it is not sufficient to collect just a few hard events to investigate the nature of zeptospace, and the reason lies in quantum mechanics. We are familiar with the concept that a physics experiment performed many times under identical conditions should give the same result, within experimental errors. Students in a laboratory never find the same result within errors, but this is a shortcoming of the students and not of the laws of physics. However, in quantum mechanics – for the revenge of many frustrated students – the familiar concept that identical experiments give identical results is simply wrong. The world of quantum mechanics is not deterministic, but is ruled by laws of probability. As the physicist Max Born once explained: "If Gessler had ordered William Tell to shoot a hydrogen atom off his son's head by means of an α particle and had given him the best laboratory instruments in the world instead of a cross-bow, Tell's skill would have availed him nothing. Hit or miss would have been a matter of chance."[4]

The result of an experiment is like rolling some dice: we cannot predict the outcome with certainty. This does not mean that the world of quantum mechanics is indecipherable. Although it is impossible to predict the result of a single experiment, quantum mechanics allows us to compute the probability of any possible outcome, which can be compared with data from an experiment repeated many times. Similarly, although we cannot foretell the single roll of a die, we can be sure that each number will come out with a probability of 1/6.

The laws of quantum mechanics govern the particle world. Therefore physicists have to collect a large number of hard events before they can make sense of the uncharted territory of zeptospace. Moreover, many of the interesting phenomena that physicists suspect are lurking in zeptospace have very low probabilities of occurring. A large number of events must be scrutinized before experiments can discover or disprove the existence of such phenomena and the validity of speculative new theories.

This is the conundrum: in order to explore zeptospace we need to record a huge number of events; but it is simply impossible to store the corresponding amount of data in any conceivable digital memory system. The problem faced by physicists is very similar to the one of a child who is obsessed by the desire of owning the Einstein miniature figure – out of the "Famous Physicists" series – contained as a gift in some of the boxes of a famous cereal brand. The Einstein figurine is very rare: on average it is found only in one box in more than a billion. Our maniac collector is very determined to find the statuette. He could buy some hundreds of cereal boxes, bring them home and check if he has been lucky. Obviously, the chance of finding Einstein would be only about one in ten million – too low to give any substantial hope. The alternative is to buy many billions of cereal boxes. Now he could be reasonably confident of finding the precious statuette, but the problem is that, even after disposing of all the furniture, his house wouldn't be large enough to accommodate all the boxes and allow a careful search for Einstein in the heap of cornflakes.

Our resourceful collector has found the solution. He buys billions of cereal boxes and, as they are carted towards his house, he stands at a street corner, near the garbage dump. Since the gifts in most of the boxes are soft toys, he quickly squeezes each box. If he cannot feel anything hard inside, he just tosses it in the dump. Only relatively few cereal boxes survives this preliminary screening and are brought to

[4] M. Born, as quoted in A. Eddington, *New Pathways in Science,* Cambridge University Press, Cambridge 1935.

his house. Later, the child will have the time to inspect the boxes stored in his bedroom at his leisure, and look for the craved Einstein figurine.

Physicists call this procedure *trigger*. The trigger is a sieve of all collision events, which selects only those that have the right characteristics to be potentially interesting. "Hard events", exhibiting tracks of very energetic particles moving off the beam direction, are kept by the trigger for a more careful analysis at a later stage. "Soft events", where most of the radiation is along the beam direction, are discarded. In order not to saturate the data storage capabilities of the analysis process, the trigger has to be extremely selective. But, because of the very intense collision rate at the LHC, the trigger decisions have to be made very quickly. If after 25 nanoseconds a decision is not taken, it is too late: another proton bunch has already collided and new events have been produced. However, in 25 nanoseconds light travels about 7 m or, in other words, much less distance than the detector size. This means that the trigger in 25 nanoseconds does not even have the time to gather information from the different parts of the detector, let alone make a decision based on that information.

The problem is daunting: it is like your boss asking you to take decisions on documents he keeps on sending you at a rate faster than you can read. The backlog of documents will hopelessly grow on your desk. To understand better how the LHC trigger circumvents this problem, let us consider another analogy.

A certain country has just introduced the new regulation that vehicles carrying more than one passenger should drive in a special lane. A guard has been assigned to stand at the border of the country and direct foreign drivers to their respective lanes. The problem is that traffic authorities have neglected to pass any regulation on speed limits and all foreign drivers enter the country at dazzling speeds. The dutiful guard is at a loss. He simply doesn't have the time to look inside all the cars, count the number of passengers, and direct them accordingly. Cars zip by before he can react. Suddenly the guard hits upon a brilliant idea. He decides to build a kilometre-long tunnel, through which all cars are compelled to drive as soon as they enter the country. Many video cameras are placed all along the tunnel, a short distance apart. Agents of his squad constantly monitor the images from these video cameras. Even though cars proceed through the tunnel at high speed, the agents have sufficient time to recognize the number of passengers and send the information to their chief, who is standing at the exit. As soon as a car emerges from the tunnel, it is immediately directed towards the appropriate lane, and his task is accomplished.

The triggers used by experiments at the LHC work similarly. Data from all events produced at the LHC are inserted in a pipeline where they are selected by a network of processors working in parallel.

Although data from a new bunch crossing comes in every 25 nanoseconds, events are analysed in the pipeline for a few thousands of nanoseconds, and only then is the on-line decision taken. This procedure requires a perfect synchronization at the level of nanoseconds for all parts of the giant detectors, or else information from different events can overlap and be hopelessly confused. The trigger actually follows a complex and elaborate architecture, with many levels of event rejection, and its procedure is different in ATLAS and CMS. At the end of this process only one event out of several million is retained for permanent storage.

In spite of this brutal reduction in the number of events carried out by the trigger, the amount of data coming from the LHC is still awesomely large. Every year the LHC produces about 10 million giga-bytes of data to be analysed offline. If one year of data were stored in CDs, the stack of disks would be more than five times taller than Mont Blanc. The computing challenge at the LHC is unprecedented.

To deal with this exceptional computing problem, the LHC experi-ments are relying on GRID-based information technology. The GRID is the next step after the World Wide Web. While the Web is a system where information is shared, the GRID distributes computer power and data storage too. Data from the LHC is shared by many computer centres around the world, connected in a large global network that works as a single computational resource. This network is based on a structure of *tiers*, where about 100 000 processors installed in 140 computer centres in 35 countries participate by providing services at different tier levels. CERN is Tier-0, where data are originally produced, stored and distrib-uted to the next level. Twelve Tier-1 institutions are responsible for long-term data storage with multiple backups and for computing analysis, while about a hundred Tier-2 institutions provide additional computing power and temporary storage service. Data transfer has to be very fast. On 15 February 2006, the system was tested with a continuous flow of physics data among institutions in Europe, Asia, and America at a rate of 1 gigabyte per second – like downloading a full-length DVD movie in 4 seconds. However, the system is planned to sustain transmission at 10 gigabytes per second.

Individual scientists of the experimental collaborations from anywhere in the world have access to the complete LHC data set recorded by their detector and they can analyse it by using computing resources from a different institution. The GRID allows a more efficient use of the global computing power, since it can manage the capacity of all computers belonging to the network. It also ensures a more robust structure, because a failure on a single node does not preclude good performance of the whole system.

The GRID is being developed not only for particle physics, but also for all scientific fields that require large computing capabilities, such as astronomy, climatology, biology, and many others. It is opening up new perspectives in international collaborations by permitting computing projects that would be impossible for a single institution. Even developing countries can find opportunities to participate in scientific projects that would be otherwise inaccessible with more limited resources.

The extreme requirements needed by the scientific community and its flexibility in adopting new technologies are strong drives towards innovation. Large scientific projects provide the perfect testing ground for developing and testing new information technologies that later become part of our everyday life. The Web is the perfect example: invented at CERN in 1989 to allow exchange of information among physicists around the globe, it was released in 1993 to public use at no cost. The GRID could become the next revolution in information technology to affect our everyday lives.

Today we are accustomed to obtaining any sort of information from the Web, in the same way as we get electricity or water in our houses without worrying what actually happens behind the electrical outlet or the water tap. A reliable global distribution system turns out to be more efficient and robust than having a power plant or a water reservoir in each house's basement. On the other hand, computing facilities are essentially local, at present. This implies that individuals and companies have to bear the effort and the cost of installing, maintaining, fixing and upgrading their own computing systems. Moreover computing resources are used very inefficiently worldwide, often either underexploited or not sufficient for the purpose.

The vision of the GRID is to make computing power a utility available to everybody – like electricity or water – in a robust system in which the user does not have to worry about where the power comes from. The GRID will be computing power out of a wall socket. There are many clear advantages and economic benefits that such a system would entail. Any small company or individual could undertake tasks impossible today without large computer facilities. CERN, developing the LHC computing grid, is playing a leading role in making this vision a reality.

Other experiments

No one believes an hypothesis except its originator but everybody believes an experiment except the experimenter.

William Beveridge[5]

---

[5] W.I.B. Beveridge, *The Art of Scientific Investigation,* Heinemann, London 1950.

The two counter-rotating proton beams intersect each other four times inside the LHC tunnel. A detector is placed at each intersection point. Besides ATLAS and CMS, the LHC hosts two other large detectors, located in caverns that predate the LHC and were used for LEP experiments. These two detectors are: A Large Ion Collider Experiment (ALICE) and the Large Hadron Collider Beauty experiment (LHCb). This book mostly concerns the exploration of zeptospace, which is the main goal of ATLAS and CMS. The goals of ALICE and LHCb are somewhat different, although their physics programmes are also interesting and, in many respects, complementary to those of the "general-purpose" detectors. ALICE and LHCb attest to the versatility of the LHC project and how different research strategies can benefit from the same machine.

During certain periods of time the LHC operates with beams of lead nuclei (and possibly other kinds of heavy nuclei) instead of protons. ALICE has been designed to study in detail such collisions. Heavy nuclei contain a large number of protons and neutrons and therefore, when they collide, many particles come into close contact, reproducing the conditions of matter at high density and temperature. According to QCD, in these conditions quarks and gluons are no longer confined inside individual hadrons, but form a new state of matter called a *quark–gluon plasma*. When two nuclei collide at the LHC, they melt, liberating quarks and gluons that can form a thermal system, creating the quark–gluon plasma. As the plasma expands, it cools down and after only about

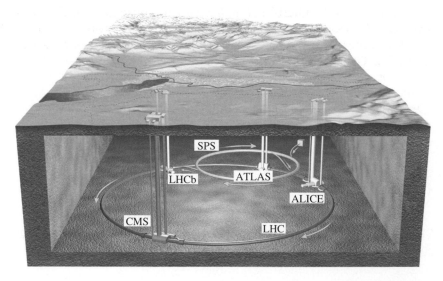

Figure 7.12 An overall view of the underground LHC tunnel with the locations of the four main detectors (ALICE, ATLAS, CMS, and LHCb).
Source: CERN.

10^{-23} seconds quarks recombine into hadrons. ALICE aims to study the properties of the quark–gluon plasma during its very short time of existence after collisions of nuclei.

The quark–gluon plasma has already been probed by experiments at RHIC (Relativistic Heavy Ion Collider) at Brookhaven National Laboratory in Upton, New York and at the CERN SPS. This new form of matter was found not to behave like an ideal gas, as was expected, but rather like a perfect fluid. It appears to be the most perfect fluid ever discovered in nature, even better than the liquid helium used to cool the LHC. The studies performed by ALICE are very interesting because they investigate the properties of QCD and the process of quark confinement. Moreover, they can teach us something about the history of the universe because the quark–gluon plasma is believed to have existed soon after the big bang, when temperature and particle densities were extremely high.

The LHCb detector, unlike ATLAS and CMS, does not completely surround the collision point, but it extends only in the forward direction along the beam. The reason is that LHCb aims at studying in detail the properties of hadrons containing the bottom quark. These states are relatively light and therefore, after being produced in a proton collision, fly mostly along the direction of the beam. The study of the bottom quark could disclose important information regarding the mystery of why quarks are replicated in three generations. It could also reveal some indirect echoes of physics occurring in zeptospace.

Finally, the LHC hosts two other smaller experiments. TOTEM (with detector components located in the CMS cavern and in the tunnel, 200 m away) has the primary goal of studying the structure of the proton. LHCf (located in the tunnel not far from the ATLAS cavern) makes measurements that are useful for the study of high-energy cosmic rays.

The human factor

If my theory of relativity is proven successful, Germany will claim me as a German and France will declare that I am a citizen of the world. Should my theory prove untrue, France will say that I am a German and Germany will declare that I am a Jew.

Albert Einstein[6]

The LHC experiments are not only pieces of equipment. Experimental collaborations of about 2500 physicists stand behind each of the two

---

[6] A. Einstein, address to the French Philosophical Society at the Sorbonne, 6 April 1922; Einstein Archive 36-378.

main detectors, and a slightly smaller number behind the other experiments put together. These people are ultimately responsible for the design, construction, and testing of every single component of these prodigious instruments. They are the people who operate the detectors and analyse the data obtained from LHC collisions, eventually announcing what has been seen in zeptospace.

The ATLAS and CMS groups are the largest international collaborations that have ever existed in particle physics. They bring together physicists from hundreds of universities and institutions, 65 countries and five continents. Detector components designed and constructed in laboratories worldwide, in the north, south, east, and west of our planet, are now tightly joined together in a single instrument, underground across the border between two European countries.

The absorber for the CMS hadron calorimeter is made of brass obtained by melting over a million decommissioned artillery shells from battleships of the Soviet Navy. It is certainly very symbolic that weapons have been melted to serve science. One day, in the ATLAS cavern, Israeli, Japanese, and Chinese physicists were mounting the thin gap chambers of the muon spectrometer that they had designed and built together. In the meantime, Russian and Polish technicians were connecting the cabling system while some Brazilian students were participating in the installation of some apparatus. Language and cultural barriers are not insurmountable.

Working in an international scientific environment is a marvellous experience. In science you are judged for your creativity and for your contributions, irrespective of your age, creed, gender, or race. Working in science teaches you to think beyond prejudice, to seek confrontation based on reasoning, to respect and accept the evidence of truth. This does not mean that scientists are perfect individuals – they have their virtues and flaws, like every other human being. However, certain values emerge in the scientific environment just as in the process of natural selection. Inability to accept evidence, blind faith in preconceptions, servitude to authority, racism, and discrimination just do not help to solve a complicated equation or to understand why a detector component does not work. Although science does not heal human weakness or malevolence, it has a natural tendency to elevate certain principles in the minds of the men and women who practise it, and to make them cultivate certain values. It is just a lucky coincidence that these principles and these values happen to be those that make a society more respectful, more honest, more just.

Science has a value for society that goes well beyond its technological innovations and its intellectual discoveries.

PART THREE
MISSIONS IN ZEPTOSPACE

8
Breaking Symmetries

Symmetry is the enemy of art.

George Bernard Shaw[1]

As we are reminded at the beginning of every Star Trek episode, the mission of the Starship Enterprise is "to explore strange new worlds, to seek out new life and new civilizations, to boldly go where no man has gone before." The mission of the LHC – the starship of zeptospace – is no less ambitious but, as you will see in the following, it may sometimes seem stranger than science fiction.

The first mission at the top of the LHC priority list is the search for a still undiscovered and rather mysterious particle: the *Higgs boson*. This chapter outlines the significance of this mission and how it can be accomplished. The issues involved in the Higgs story are rather theoretical and some of its aspects are quite technical. Their explanation will then require us to follow a rather long logical path. We will start by exploring the fundamental role of symmetries in physics, continue with the concept of spontaneous symmetry breaking, and finally reach the Higgs boson and the hunt for it at the LHC.

Symmetry and mathematics

Mathematics may be defined as the subject in which we never know what we are talking about, nor whether what we are saying is true.

Bertrand Russell[2]

Mathematics is the language of nature. Each time a new physical phenomenon is understood, it is invariably described in mathematical

---

[1] G.B. Shaw, as quoted in M. Holroyd, *Bernard Shaw: The Lure of Fantasy*, Random House, New York 1991.

[2] B. Russell, *Recent Work on the Principles of Mathematics*, in *International Monthly*, vol. 4 (1901).

terms. The observation of natural processes reveals regularities that can be ascribed to physical laws, expressible by mathematical equations. Knowledge of nature does not mean knowledge of the exact positions and velocities of every atom in the universe at every instant, but rather the identification of the fundamental laws that determine the behaviour of the physical world. It is remarkable how simple these laws are, in spite of the complexity of the natural phenomena that we observe.

The mathematical formulation of the physical laws often exposes unsuspected connections between different aspects of a phenomenon and even between theories that describe different phenomena. For instance, the unification between electricity and magnetism becomes evident only after observing the mathematical structure of Maxwell's equations. Mathematics is the only language known to us capable of describing these connections in an exact manner, free of any ambiguity.

The role of mathematics as the language of nature became evident as soon as modern science began. Galileo wrote in 1623: "Philosophy is written in this grand book (I mean the universe) which stands continually open to our gaze, but it cannot be understood unless one first learns to comprehend the language and to interpret the characters in which it is written. It is written in the language of mathematics, and its characters are triangles, circles, and other geometrical figures, without which it is humanly impossible to understand a single word of it; without these, one is wandering about in a dark labyrinth."[3]

The physicist Eugene Wigner speaks of "the unreasonable effectiveness of mathematics," asserting that "the enormous usefulness of mathematics in the natural sciences is something bordering on the mysterious and there is no rational explanation for it."[4] The choice of nature to use a mathematical language is the key to its intelligibility. But the primary reason for the existence of physical laws is still unexplained. Could there exist a universe without physical laws? "The eternal mystery of the world is its comprehensibility,"[5] wrote Einstein. The most incomprehensible aspect of nature is that we can comprehend it.

The fundamental physical laws look simple to us only after we have been able to express them in the appropriate language, which often involves the use of advanced mathematics. "The simplicities of natural laws arise through the complexities of the languages we use for their expression,"[6] as Wigner stated. Very often abstract mathematical

[3] G. Galilei, *The Assayer (Il Saggiatore)*, 1623.

[4] E.P. Wigner, *The Unreasonable Effectiveness of Mathematics in the Natural Sciences,* reprinted in *The Collected Works of Eugene Paul Wigner,* vol. VI, ed. J. Mehra, Springer-Verlag, Berlino 1995.

[5] A. Einstein, *Physics and Reality*, The Journal of the Franklin Institute, vol. 221 no. 3 (1936), reprinted in A. Einstein, *Ideas and Opinions*, Crown Publishers, 1954.

[6] E.P. Wigner, Communications in Pure and Applied Mathematics 13, 1 (1959).

structures, invented for the sake of pure speculation, turn out to become useful, if not indispensable, to describe certain natural phenomena and the physical laws they obey. Calculus was necessary for the development of classical mechanics, non-Euclidean geometry for general relativity, complex analysis for quantum mechanics, group theory for particle physics, and even more sophisticated mathematical structures find applications in modern string theory. *Symmetry* is a mathematical concept which has now become one of the primary elements of physics. If mathematics is the language of nature, symmetry is its syntax.

We usually refer to symmetry as a correspondence, visibly perceptible, between different parts of a system. But it is also natural to associate with the notion of symmetry the idea of balance, beauty, and harmony among proportions. The numerical ratios of the ideal human body, expressed by the Kanon of Polykleitos, were synonymous with perfection and symmetry for all sculptors of the classical period. Even Goethe agrees with this point of view: "By the word symmetry...one thinks of an external relationship between pleasing parts of a whole,...a strength out of weakness, a beauty out of ordinariness."[7]

The concept of symmetry has a precise mathematical definition, which corresponds to the *invariance of a system under a transformation*. In simpler words, symmetry exists when a system does not change under some well-defined manipulation. A circle has rotational symmetry because, when rotated about its centre, it remains the same. A square remains invariant only if we rotate it by an angle of 90 degrees or by one of its multiples. The circle is said to have a *continuous symmetry*, because it is invariant under a transformation that can be made infinitesimally small. Instead, if a transformation corresponds to an abrupt variation, which cannot be made infinitesimally small, we speak of *discrete symmetry*. The square has only a discrete symmetry because invariance is maintained only when the rotation angle is of 90 degrees or multiples thereof. Invariance under mirror reflection is another example of a discrete symmetry.

In physics, the concept of symmetry complies with the mathematical definition. And yet the more vague connotation of symmetry, associated with the idea of beauty and harmony, also plays a compelling role. Nature seems to take pleasure in exploiting all possible symmetries for her fundamental laws, like a painter eager to use all the most splendid colours on her palette. Physicists take advantage of this tendency of nature towards symmetry, using it as a hint to deduce the properties of the particle world.

The use of symmetry in modern mathematics started with the tones of a romantic tragedy in the Parisian night of 29 May 1832. A young

---

[7] J.W.V. Goethe, *Naturwissenschaftliche Schriften,* in *Collected Works*, Princeton University Press, Princeton 2006.

man, only 20 years old, feverishly jolts some confused scribbles, nearly illegible, by the light of a candle. He is writing some mathematical notes but, among the equations, some words indicate an approaching calamity: "une femme," "je n'ai pas le temps," "Stéphanie." At dawn, he is found at the edge of a country road with a bullet in his abdomen.

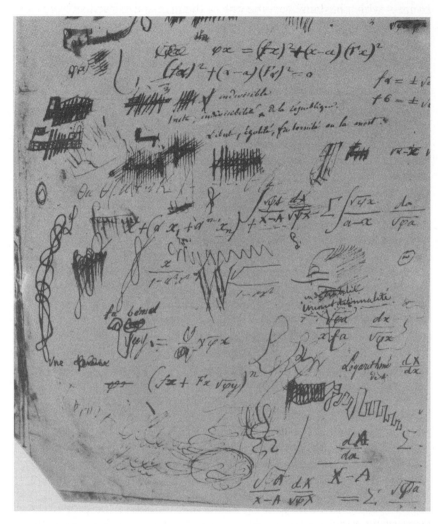

Figure 8.1 One of the pages left on the desk of Évariste Galois before he left for his fatal duel. Visible are the words *"Une femme"* (scribbled out at the bottom left corner), *"Pas l'ombre"* (top left corner), and the republican motto *"Liberté, égalité, fraternité ou la mort"*.

Source: R. Bourgne and J.P. Azra, *Écrits et mémoires mathématique d'Évariste Galois*, Gauthier-Villars, Paris 1962.

That young man was Évariste Galois (1811–1832), a mathematical genius unrecognized during his short existence. In the turbulent epoch between the end of the Napoleonic period and the restoration of the monarchy, his exuberant republican fervour pushed him to several acts of rebellion and, consequently, to arrest and incarceration. During an outbreak of cholera in Paris, the youngest among the prisoners were moved to the Pension Sieur Faultrier, where Galois fell in love with the daughter of a resident physician, Stéphanie-Félicie Poterine du Motel. Only a month after his release, he was killed in a pistol duel.

The motive for the duel remains an unsolved mystery, although there is no lack of romantic interpretations. The writer Alexandre Dumas in his *Mes Memoirs* maintains that Stéphanie's fiancé, Pescheux d'Herbinville, had discovered the love affair and challenged Galois to a duel. D'Herbinville was known to be one of the best pistols in France, and Galois, knowing that he had no way out, hurried to write his scientific will in a single night. Eric Bell, in his classic book *Man of Mathematics*, conjectures that the duel was staged by the secret police to eliminate an active member of the revolutionary republican movement. According to this interpretation, Stéphanie was a shrewd temptress and d'Herbinville an agent of the royal police. Others suggested that Galois himself plotted his death together with his friends of the Société des Amis du Peuple, a republican group, so that his funeral could provide the spark to ignite the revolution.

The story of the fateful events preceding Galois' death rests forever enshrouded in myth, but his work, rewritten and interpreted in 1846 by the great mathematician Joseph Liouville, remained a source of inspiration for generations of scientists. Galois tried to identify the solutions of certain algebraic equations, which involve the fifth or higher powers of the variable x. The most interesting aspect of Galois' work was the introduction of new conceptual structures – the *groups* – that describe permutations among the different solutions of the algebraic equations. By decomposing these groups, he identified the fundamental elements of permutations. Just as particles are the building blocks of matter, the groups identified by Galois are the building blocks of symmetry. The use of symmetry gave Galois a completely new perspective and allowed him to extract properties of the solutions of algebraic equations that cannot be identified by other methods.

The next step in the understanding of the mathematical properties of symmetries required an energetic man. Sophus Lie (1842–1899) fitted the role perfectly. Born in a small village in Norway, Lie was sent to school in Oslo after his mother's death. When he wanted to go back home, young Sophus covered the 50 km journey by foot. The story goes that once, after realizing that he had forgotten a book at home, he walked back and forth in the same day.

Lie's beginnings in mathematics were slow but, when 27 years old, he won a grant that allowed him to move to Berlin, where he found a very fertile academic environment. In 1870, he visited Paris, where he discussed the theory of permutation groups with the mathematician Camille Jordan. When the French–Prussian war broke out, he left Paris just in time to escape the terrible siege of the 1870 winter. Not unusually for him, he started by foot for Milan, in Italy, where he wanted to meet the mathematician Luigi Cremona, an expert in algebraic curves and surfaces. In the meantime, the army of Napoleon III was besieged at Metz. Lie, a massive long-bearded man marching alone across the French countryside, attracted attention and was stopped by the police. The mathematical notes found in his baggage were believed to be a secret dispatch written in code with mysterious symbols. Without appreciating the difference between a Norwegian and a German accent, the police arrested him on suspicion of being a spy in the Prussian army. He was freed after one month in prison, only on account of the intervention of his friend, the French mathematician Jean-Gaston Darboux.

Lie had great admiration for Galois' work and he wanted to treat differential equations the way Galois had treated algebraic equations. Algebraic equations have a finite number of solutions, at most equal to the largest power of the variable x. The transformations studied by Galois describe the operations of exchange between different solutions. These are abrupt variations, which therefore correspond to discrete symmetries. Differential equations, by contrast, describe smooth variations and have an infinite number of solutions. Their classification would then lead to the discovery of the fundamental elements of continuous symmetries.

There is a deep connection between continuous symmetries and geometrical structures. For instance, the rotational symmetry on a plane identifies the circle, because the circle is the only (simply connected) geometrical form perfectly invariant under rotations about its centre. In the same way, spatial rotational symmetry identifies the sphere. Lie started to classify all possible geometrical structures associated with operations of continuous symmetry, a work later completed by Wilhelm Killing and Élie Cartan. These geometrical structures are too complicated to be visualized in our three-dimensional space, but can be perfectly well described by the language of mathematics.

In this way, Lie identified the building blocks of continuous symmetries, which are today called *Lie groups*. The Lie groups are the abstract entities that summarize the essence of continuous symmetries. As is often the case in mathematics, the structures discovered by Lie transcend the specific problem that he was considering. Their properties are universal and are valid beyond differential equations or spatial rotations. Just as every poem, no matter how sublime, is ultimately a combination of letters, so a continuous symmetry, no matter the context in which it is realized, is a combination of Lie groups. The symmetries with which nature enriches the particle world are no exception, and they can be formulated in terms of Lie groups as well.

Symmetry and physics

> Tell me why should symmetry be of importance!
>
> Mao Tse-tung[8]

In spite of the progress made in mathematics, symmetries did not play a big role in classical physics and their use had been limited to some applications in crystallography. The change of pace began with Einstein. The starting point of special relativity is the assertion that the speed of light is the same for different observers or, more precisely, that it is invariant under the change of reference frames moving with uniform relative velocities. Einstein had elevated invariance to the status of a fundamental principle. But invariance is synonymous with symmetry, and thus the principle is that symmetry dictates the physical laws. This was a decisive departure from tradition, because in classical physics it had always been the other way around: the physical laws are the primary elements, which then determine the symmetries.

This new approach was a real conceptual revolution, and even a genius like Lorentz had difficulty at the beginning in accepting this point of view. In some lecture notes dating from 1906, he makes some begrudging comments: "Einstein simply postulates what we have deduced, with some difficulty and not altogether satisfactorily, from the fundamental equations of the electromagnetic field."[9] Lorentz had tried to derive the constancy of the speed of light from the laws of electromagnetism, but his results were most unsatisfactory. Nevertheless it seemed to him that Einstein was evading the question, by just postulating the answer.

But the idea that symmetries determine physical laws and not the converse turned out to be the winning choice. It provided the inspiration for Einstein's formulation of general relativity, a theory that is completely determined by the condition of invariance under arbitrary motion of the observer. The existence of gravity is a mere consequence of an invariance principle or, in other words, of symmetry.

The role of symmetries in physics has grown ever since. A very important step in this direction was made by Emmy Noether (1882–1935), a mathematician who made fundamental contributions to the field of abstract algebraic structures. Among physicists she is best

---

[8] M. Tse-tung, as quoted in C.C. Gaither and A.E. Cavazos-Gaither, *Mathematically Speaking*, Institute of Physics Publishing, Bristol 1998.

[9] H.A. Lorentz, *The Theory of Electrons and its Applications to the Phenomena of Light and Radiant Heat,* Dover Publications, New York 1952.

known for the so-called *Noether's theorem*, which clarified the deep connection between continuous symmetries and conservation laws. Conservation laws were old acquaintances from classical physics. They state that certain measurable quantities – like energy, momentum, or electric charge – do not change in physical processes. Noether's theorem asserts that any continuous symmetry leads to a conservation law.

It is rather intuitive that a conservation law must lurk behind any continuous symmetry. After all, symmetry reflects invariance under a transformation, and therefore there must exist a quantity that remains invariant or, in other words, that is conserved. For instance, a circle is invariant under rotations about its centre. The various points of the circle move during the rotation, but their distance from the centre remains invariant. Hence, the symmetry of the circle is associated with the conservation of the distance between each of its points and the centre.

The power of Noether's theorem was to show that this intuitive concept is valid for any continuous symmetry, not just for geometrical transformations, but also for the more abstract transformations considered in particle physics. For instance, from Noether's theorem we discover that the conservation of electric charge is the consequence of the special rotational symmetry of QED. This rotation does not act on the physical space, but on an abstract space defined by the quantum fields.

Emmy Noether now has a place in the history of science, but a university career for a woman was not easy in her time. In 1915 the University of Göttingen refused her "habilitation" on the basis of her gender. "What will our soldiers think," barked one faculty member, "when they return to the university and find that they are required to learn at the feet of a woman?"[10] The famous mathematician David Hilbert, appalled by the decision, declared in front of the Academic Senate: "I do not see that the sex of the candidate is an argument against her admission as Privatdozent. After all, we are a university and not a bathing establishment."[11] The University granted her "habilitation" only in 1919, in the more liberal atmosphere of the Weimar Republic.

Noether had a unique passion for mathematics and she was always surrounded by a large group of students and assistants with whom she discussed with great fervour and animation. When it became compulsory to teach only "Aryan mathematics" in Germany – a subject in which she was evidently ignorant – Emmy Noether was expelled by the university. In 1933 she succeeded in emigrating to the USA, where she declared to have spent the happiest period of her life. But only a year

---

[10] L.M. Osen, *Women in Mathematics*, MIT Press, Boston 1974.
[11] H. Weyl, *Scripta Mathematica*, 3, 201 (1935).

and a half later, she died after tumour-related surgery. In her obituary, Einstein wrote: "In the judgement of the most competent living mathematicians, Fräulein Noether was the most significant creative mathematical genius thus far produced since the higher education of women began."[12]

Gauge symmetry

> We are all agreed that your theory is crazy. The question which divides us is whether it is crazy enough to have a chance of being correct.
>
> Niels Bohr[13]

A geometrical symmetry can be visually perceived, but the concept of symmetry among particles sounds somewhat abstract. What do we mean by saying that a theory of particle physics exhibits symmetry? Let me start with an example. In Chapter 3 it was mentioned that protons and neutrons respond to the strong force in a nearly identical manner. The "nearly" refers to small differences that can be imputed to the effect of the electromagnetic forces felt by the charged proton but not by the neutron. Making the idealization of neglecting electromagnetism with respect to the strong force – which is a fairly good approximation for the real world – one finds that the result of any physical process remains the same once we replace all protons with neutrons and vice versa. Said in a more elegant way: the strong force is symmetric under exchange of protons and neutrons.

At this point, one would think that the symmetry is *discrete*. After all, exchanging protons with neutrons is an abrupt operation, typical of discrete symmetries. But, once again, our intuition is defeated by the odd world of quantum mechanics: in fact, the symmetry of the strong force is *continuous*. Just as the Egyptian god Sekhmet was represented in various forms combining human and lion elements, so in quantum mechanics particles can exist in hybrid states, which are neither protons nor neutrons, but their combination. We cannot tell with certainty whether an experiment measuring such states will detect a proton or a neutron. Quantum mechanics does not describe a deterministic world and only assigns certain probabilities for the experiment to detect a proton or a neutron.

We can imagine the state of a particle as a knob that controls the volume on a radio. The knob can be turned continuously from MIN to

---

[12] A. Einstein, *The New York Times*, 5 May 1935.

[13] N. Bohr to W. Pauli, in *Symposium on Basic Research,* ed. D. Wolfle, American Association for the Advancement of Science, Washington 1959.

MAX and can be adjusted to any intermediate position. Similarly, the state of a particle can be (metaphorically) rotated between proton and neutron, going though an infinite number of intermediate hybrid states. The symmetry of the strong force corresponds to the invariance of its laws under a continuous rotation of the particle states.

This example illustrates how there can exist continuous transformations among particles (or, more precisely, among fields). These transformations are completely analogous to familiar rotations but, instead of taking place in ordinary space, they involve an abstract space where points are actually particle states. But this is an irrelevant difference for the mathematics, and the same Lie groups that represent spatial rotations and their generalizations describe transformations among particles too. Just as a geometric figure can be symmetric (that is invariant under a transformation of points in space), so a theory of the subatomic world can also be symmetric (invariant under a transformation of particles).

An important characteristic in the previous example of protons and neutrons is that the transformation of particles must occur simultaneously in all points of space. Indeed, the invariance of physical processes holds only if the role of protons and neutrons is exchanged *everywhere* in space. Symmetries of this kind are called *global*. A global symmetry, in which particles are simultaneously transformed everywhere in space, reminds us of action at a distance – a concept completely incompatible with special relativity. Hence, it is no surprise that nature seems to show very little respect for global symmetries that, as far as we know, are not part of the fundamental principles of the particle world.

It is natural then to wonder whether a theory of particle physics could possess a *local symmetry* – a symmetry associated with transformations among particles acting differently from point to point in space and time. Local, as opposed to global, symmetries have the right features to become essential ingredients of a fundamental theory of matter and forces. I suppose that this is what Naomi Klein, the anti-globalization activist, had in mind when she declared in an interview: "So a tension developed between global and local.... The global was becoming increasingly abstract."[14]

The adjective "local" may give the impression of a more restricted concept with respect to "global". Instead, the existence of a local symmetry is a very demanding requirement on a theory: the physical laws have to remain invariant even when the transformation acts differently from point to point. For instance, a circle is certainly not invariant under a *local* rotation, that is, under a transformation in which each point of the circle is rotated by a different angle. Such a transformation

---

[14] M. Chihara, *Naomi Klein Gets Global*, Global Policy Forum News, 25 September 2002.

could completely destroy the circle, and could change it into a semicircle or even into a single point. A system must contain an additional element in order to exhibit local symmetry. The concept is undoubtedly rather abstract and an example can help to clarify this idea.

In summer a field of bright yellow sunflowers offers a spectacular view. In the morning all the faces of the flowers are turned towards the east. As the day goes by, the flowers slowly turn their faces westwards in perfect synchronization. The property of "facing the same direction" remains unchanged as the flowers rotate. This is an example of a *global* symmetry, because the invariance holds only if the rotation is identical for every flower at every given moment.

Imagine a field of unruly sunflowers where, irrespective of the position of the sun, each flower constantly changes its orientation during the day, turning its face towards a random direction. The rotation of the flowers is now a *local* transformation because it is different from flower to flower. Obviously the property of "facing the same direction" is not respected by this local transformation.

A diligent farmer decides to re-establish order in his field of rebellious sunflowers. He plants each flower inside a pot buried in the ground. Each pot is supplied with a special electronic monitor. Whenever a sunflower turns its face, its pot counter-rotates by an equal and opposite angle, thus undoing the original rotation. Thanks to the system of "rotating pots" all the sunflowers are carefully kept in the same orientation, in spite of their tendency to turn haphazardly. The property of "facing the same direction" has now a local symmetry, because it holds even if each individual flower chooses to turn its face independently of all the others.

The moral of this metaphor is that a *global* symmetry can be promoted into a *local* symmetry by introducing a new element (the "rotating pots"). In the language of particle physics, this new element is called a *gauge field* and the local symmetry is called a *gauge symmetry*.

The gauge field is like an elastic fabric that stretches everywhere by exactly the right amount to compensate for the change of any element of the system, just as the "rotating pots" compensate the individual rotations of the sunflowers. The self-adapting fabric of the gauge field is able to properly readjust itself at every point of space and at every instant in time in order to ensure the invariance of the system and preserve the gauge symmetry.

The most remarkable aspect of the story is that this elastic fabric is not just an abstract invention of some deranged theoretical physicist, but it turns out in fact to be a very real, very concrete entity. The gauge field is nothing other than the electromagnetic field describing the photon.

The punch line is that the electromagnetic force is just the consequence of a gauge symmetry of nature. Once the symmetry principle has been enunciated, the interactions between particles are completely determined, and the existence of the photon is an inescapable consequence. This is the conceptual revolution of the gauge principle: not force, but symmetry is the primeval notion. In the words of Cheng-Ning Yang: "Symmetry dictates interaction."[15]

Not only electromagnetism, but also gravity is dictated by a symmetry principle. General relativity has taught us that force is not the starting concept. Properties of space-time – also linked to a gauge symmetry principle – determine the structure of the theory; the gravitational force is only a consequence. All forces must obey the laws of symmetry. "Symmetries are laws, which the laws of nature have to observe,"[16] as concisely expressed by Wigner.

The symmetry at the origin of electromagnetism (or, equivalently, of QED) corresponds to the simplest element in the classification by Lie: it is equivalent to the rotation symmetry of a circle. But Lie put into our hands the bricks to build any kind of continuous symmetry, even those that we cannot visualize in three-dimensional space. What happens when we consider gauge symmetries based on more complicated Lie groups than the one of QED?

This is the question that two young researchers, the Chinese-born Cheng-Ning Yang (Nobel Prize 1957) and the American Robert Mills (1927–1999), decided to address. Yang recalls: "In 1953–1954, I was visiting Brookhaven and Bob [Mills] was my office mate. We discussed many things in physics…It was in that year that we found the very elegant and unique generalization of Maxwell's equations. We were pleased by the beauty of the generalization, but neither of us had anticipated its great impact on physics 20 years later."[17]

Yang and Mills discovered a generalization of QED based on a gauge symmetry described by an arbitrary Lie group. The structure of the theory is very simple and elegant. The main difference with QED lies in the number of gauge fields. While in QED there is a single gauge field (the photon) transmitting the electromagnetic force, more elaborate symmetry structures lead to theories with many "photons". These "photons" – the gauge particles – are the carriers of the gauge force described by the theory. In contrast with the case of QED, the "photons" of the Yang–Mills theory interact with each other. Because of this property, Yang and Mills were hoping to use their theory to explain the strong force and to describe Yukawa's pion.

[15] C.N. Yang, *Selected Papers 1945–1980 with Commentary*, Freeman, San Francisco 1983.

[16] E.P. Wigner, *The Collected Works of Eugene Paul Wigner*, vol. VI, ed. J. Mehra, Springer-Verlag, Berlino 1995.

[17] C.N. Yang, in *50 Years of Yang–Mills Theory*, ed. G. 't Hooft, World Scientific, Singapore 2005.

Sadly, the hope was ill-founded and the debut of the theory was not very successful. In February 1954, Oppenheimer invited Yang to give a seminar at Princeton, and this is Yang's recollection: "Pauli was spending the year in Princeton, and was deeply interested in symmetries and interactions. Soon after my seminar began, when I had written down on the blackboard my first equation..., Pauli asked, 'What is the mass of this field?' I said we did not know. Then I resumed my presentation, but soon Pauli asked the same question again. I said something to the effect that that was a very complicated problem, we had worked on it and had come to no definite conclusions. I still remember his repartee: 'That is not sufficient excuse'. I was so taken aback that I decided, after a few moments' hesitation to sit down. There was general embarrassment. Finally Oppenheimer said, 'We should let Frank [Yang] proceed'. I then resumed, and Pauli did not ask any more questions during the seminar. I don't remember what happened at the end of the seminar. But the next day I found the following message: 'February 24. Dear Yang, I regret that you made it almost impossible for me to talk with you after the seminar. All good wishes. Sincerely yours, W. Pauli.' "[18]

Pauli had immediately pointed out what appeared to be the Achilles' heel of the Yang–Mills theory. He had asked: "What is the mass of the particle corresponding to the gauge field?" This question really posed a massive problem.

A massive problem

> No problem is too big to run away from.
> Charles Schultz[19]

Today we know that electromagnetic, weak, and strong forces are described by the Standard Model, which is a gauge theory of the kind invented by Yang and Mills. This wisdom is the fruit of great experimental discoveries and fundamental theoretical breakthroughs, some of which were briefly outlined in Chapter 4. Behind the Standard Model lies the profound principle that all known forces are the consequence of symmetry.

The abstract concept of the gauge field – the elastic fabric that stretches to ensure that symmetries are preserved in their highest form – embodies the carriers that transmit all known forces in the real world. The gluons, *W*, *Z*, and photon are the gauge particles dictated by the

---

[18] C.N. Yang, *Selected Papers 1945–1980 with Commentary*, Freeman, San Francisco 1983.

[19] C.M. Schultz, *Charlie Brown*.

symmetry of the Standard Model. The symmetry principle elegantly describes all forces of nature. And yet this idyllic picture of the particle world has a glitch.

Gauge symmetry is an extremely powerful principle for ruling over all properties of forces. But its inflexibility is the cause of a problem. Gauge symmetry decrees that all force carriers must be particles with zero mass. The photon and the gluon do indeed have zero mass, but the *W* and *Z* are known to have large masses. This is the difficulty that had haunted gauge theories since their beginnings, as emphasized by Pauli's question at Yang's seminar. How can we reconcile gauge symmetry with massive force carriers like the *W* and *Z*? This question, referred to as the *problem of electroweak symmetry breaking*, is playing a central role in research in particle physics today. The identification of the right answer to this problem is one of the main tasks of the LHC.

The irreconcilability of gauge symmetry with massive *W* and *Z* lies in a crucial distinction between force carriers with and without mass. Let me explain the difference. According to the quantum-mechanical duality between particles and waves, photons are associated with electromagnetic waves. It is a well known empirical fact that electromagnetic waves correspond to oscillations of electric and magnetic fields in the plane orthogonal to the direction in which waves propagate; no oscillations occur along the direction of motion. Waves that oscillate in directions orthogonal to the motion are called *transverse*. On the other hand, waves that oscillate along the direction of motion, like the sound waves produced by alternating air compression and decompression, are called *longitudinal*.

In general, waves contain both transverse and longitudinal components. For instance, water waves have both components, since water molecules move in circles as the wave propagates. Electromagnetic waves, however, are purely transverse. The absence of longitudinal components in electromagnetic waves is not fortuitous, but is a direct consequence of gauge symmetry. Gauge symmetry acts like a filter eliminating oscillations of the electromagnetic wave along the longitudinal component, just as Polaroid lenses in sunglasses polarize light by filtering out one of its components.

Special relativity affirms that we can never catch up with a light ray, no matter how fast we move. This statement, curious but apparently innocuous, actually has far-reaching consequences and it is at the origin of all the counterintuitive and confounding aspects of special relativity. Suppose, for instance, that you borrow from CERN's director general his HSCT – the high speed civil transport which, according to Dan Brown's *Angels and Demons*, "runs on slush hydrogen" and "goes Mach fifteen."[20] Even if you pursue a light ray with the HSCT, your

---

[20] D. Brown, *Angels and Demons*, Washington Square Press, New York 2000.

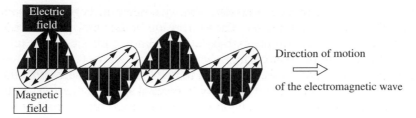

Figure 8.2 Electromagnetic waves are purely transverse, because the electric and magnetic fields oscillate only in directions perpendicular to the direction of motion.

measurement of its velocity will give exactly the same result as the measurement made by a friend of yours, who is comfortably sitting in the laboratory watching the light ray zip by. No matter how fast you move, light always travels at the same speed with respect to you. In other words, you can never observe a photon at rest.

The situation changes when, instead of the photon, we consider massive force carriers like the W and Z. Massive particles always move with velocities smaller than the speed of light and thus we can catch up with them. When a massive particle is viewed at rest, there cannot be any distinction between transverse and longitudinal components of its associated wave, because there is no direction of motion. Hence, waves corresponding to massive force carriers must contain *both* transverse and longitudinal components, just because all directions of space are equivalent. This brings in the clash with gauge symmetry. Gauge symmetry filters out longitudinal components and thus cannot describe massive force carriers. This is the essence of the problem of electroweak symmetry breaking: gauge symmetry is incompatible with massive W and Z. And yet, weak interactions appear to be adequately described by gauge symmetry.

This conceptual conflict explodes into a problem once we zoom into small distances. Even before experiments explore a new region of microscopic space, we can use our imagination and plunge into smaller distances with the help of theoretical calculations. Theoretical calculations are like a virtual starship that allows us to penetrate into the depth of space, by extrapolating our knowledge to unexplored worlds. Let us imagine travelling with this virtual starship towards smaller and smaller distances. Suddenly, once the starship crosses the entrance gate of zeptospace, an alarm for red alert goes off in the control panel. The engine has failed and we cannot advance any further. The control panel of the virtual starship communicates that the longitudinal waves of the massive W and Z interact with each other with a probability larger than 100 per cent. This is, of course, sheer nonsense, because a probability of 100 per cent means certainty and thus probabilities cannot exceed

100 per cent. This result is an indication that something is going berserk in the theoretical calculation. The conclusion is that new phenomena, not described by our simple extrapolation of known physics, *must* take place in the land of zeptospace.

This result is one the main reasons for the great excitement about the LHC. Once zeptospace is explored with the real starship (the LHC), we will find out why the virtual starship (the theoretical calculations) broke down at the entrance gate. New particles or new forces *must* exist in zeptospace to resolve the problem of electroweak symmetry breaking.

The LHC will be the indisputable and final judge handing down the verdict on the nature of the phenomena associated with the problem of electroweak symmetry breaking. Nonetheless, enough clues have been collected from previous experiments to make theoretical physicists reasonably confident that the right answer to the problem has already been guessed. And the answer is the Higgs mechanism. But to understand how the Higgs mechanism deals with the problem, we must first introduce a new concept.

Spontaneous symmetry breaking

> Like the ski resort full of girls hunting for husbands and husbands hunting for girls, the situation is not as symmetrical as it might seem.
>
> Alan Mackay[21]

Although the terminology sounds a bit arcane, the phenomenon of *spontaneous symmetry breaking* is common even in classical physics. Let us start by going back to the time of Newton. The gravitational law of the inverse of the square of distance had already been guessed at before Newton by the French notary, and amateur astronomer, Ismaël Bullialdus (Boulliau) in his *Astronomia Philolaica* of 1645 and, later, by the English physicist Robert Hooke, the archenemy of Newton. But no one before Newton had understood that such a law could explain planetary motion.

The conceptual stumbling block had a lot to do with symmetries. Consider the gravitational force exerted by the sun. If the force is radial – that is, it depends only on distance – then the system has perfect rotational symmetry around the sun and any planetary orbit should be circular. But this conclusion is in contradiction with Kepler's observation that planets follow elliptical orbits. Newton's simple but profound observation was that planets have initial velocities that do not respect

---

[21] A.L. Mackay, lecture at Birkbeck College, University of London, 1964, as quoted in A.L. Mackay, *The Harvest of a Quiet Eye*, Institute of Physics Publishing, Bristol 1977.

the rotational symmetry around the sun. Hence, elliptical orbits are not in contradiction with radial forces. Using more modern terminology, the moral is: the symmetry of an equation (the gravitational law) is not necessarily the symmetry of its solution (the planetary orbit).

Consider the story of an unfortunate individual who has been locked inside a room with no windows since birth. This poor creature develops an interest in science and wants to discover the laws of physics. From experiments conducted in his room he concludes that the vertical direction is special, because objects fall downwards. If the vertical direction is different from the others, then space is not symmetric under rotations, just as a cylinder is not rotationally invariant while a sphere is.

Ancient philosophers, although not locked in rooms since birth, fell into the same logical trap as our imaginary amateur scientist and gave a special meaning to motion along the vertical direction. It took the genius of Galileo and Newton to understand that fundamental laws do not single out any special direction, but that we just happen to live in a place subjected to the gravitational attraction of the earth. Moral: certain physical systems (the interior of the room) do not exhibit all the symmetries of the fundamental laws that govern them.

These examples show that symmetry, although present in the physical laws, need not be manifest in the particular circumstances of the system. The presence or absence of symmetry in a system can be the result of an accident of its special conditions. But the goal of fundamental physics is to identify the symmetries of the laws of nature, whether or not these symmetries are manifest in a particular system.

An especially interesting situation occurs when the system is spontaneously driven into a state that does not respect some of the symmetries of the physical laws. This phenomenon is called *spontaneous symmetry breaking*. The word "breaking" is certainly a misnomer, because symmetry, though not manifest in the state of the system, is still exactly present in the fundamental laws.

The classic example to illustrate the phenomenon of spontaneous symmetry breaking is the case of ferromagnetism. Ferromagnets are materials that form natural permanent magnets. The magnetic field generated by a ferromagnet identifies a special direction inside the material (the direction connecting "north" to "south" magnetic poles). This special direction violates rotational symmetry. But the laws of electromagnetism, which govern ferromagnets, are perfectly symmetric under rotations. We are facing here the phenomenon of spontaneous symmetry breaking.

Atoms behave like microscopic magnets because of the motion of electric charges in their interior. Usually inside a material, the atomic magnets point in random directions and they cancel each other out, without giving any macroscopic effect. But inside a ferromagnetic material, the mutual interactions among atoms drive all microscopic magnets to point in the

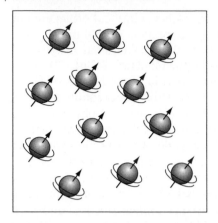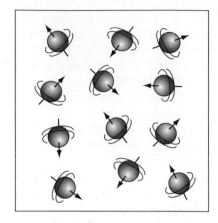

Figure 8.3 In a ferromagnet the atomic magnets are oriented all in the same direction and generate a natural magnetization of the material (left frame). Above a critical temperature, the orientations of the atomic magnets are random and the magnetization disappears (right frame).

same direction because this is the energetically favoured configuration (see Figure 8.3). This global alignment causes a natural magnetization of the material. From the point of view of energy considerations, it does not really matter which direction is selected, but only that all atoms point in the same direction. Naively one could believe that the configuration in which all atoms are aligned is more "symmetric" because it has a higher degree of order, but actually the opposite is true. The configuration in which the orientations of the atomic magnets are random is symmetric under rotations because the system remains the same when it is rotated. On the other hand, a ferromagnet looks different when it is turned, and therefore the aligned configuration inside a ferromagnet is not rotationally symmetric.

If microscopic creatures lived inside a ferromagnet, they would observe the precise alignment of atoms all around them and they would be fooled into believing that space is not rotationally symmetric. But the laws of electromagnetism *are* symmetric, although the system is spontaneously driven into a state that is *not* symmetric. This is the phenomenon of spontaneous symmetry breaking. However, when a ferromagnet is heated above a critical temperature (770°C for iron), the thermal motion randomizes the directions of the atomic magnets and the rotational symmetry is re-established in the system, while the natural magnetization disappears.

Why should spontaneous symmetry breaking be relevant for the electroweak force? The problem of electroweak symmetry breaking is the reconciliation between the two apparently conflicting concepts of gauge symmetry and massive W and Z. But, in the presence of spontaneous symmetry breaking, the notion of symmetry takes a new form. It is then necessary to reconsider gauge symmetries in the light of spontaneous symmetry breaking.

The Higgs mechanism

> A vacuum is a hell of a lot better than some of the stuff that nature
> replaces it with.
>
> Tennessee Williams[22]

The phenomenon of spontaneous symmetry breaking is very special in the case of gauge symmetry and it is called the *Higgs mechanism*, after the British physicist Peter Higgs. The essence of the Higgs mechanism is that a new quantum field, called the *Higgs field*, pervades all space. We have already learned that every particle that we observe is in reality only the manifestation of ripples in a quantum field. So, in this respect, there is nothing really new. But the peculiarity of the Higgs field is that, even after all particles are removed, space is not empty. Space is filled with a uniform distribution of Higgs field, which will be called here the *Higgs substance* (although in physics parlance it goes by the name of *Higgs vacuum expectation value*). This unusual situation occurs because the energy of space filled with Higgs substance is less than the energy of empty space. In other words, nature saves energy by filling space with Higgs substance rather than leaving it empty. Thus, Higgs substance spontaneously permeates all space, just as global magnetization is spontaneously created inside a ferromagnet. The vacuum – the configuration of minimal energy – is not just "nothing", but a medium impregnated with Higgs substance.

The magnetization in a ferromagnet identifies a special direction in space. This special direction spontaneously breaks rotational symmetry. Similarly, the Higgs substance identifies a special direction not in physical space, but in the abstract space of quantum fields. This special direction spontaneously breaks gauge symmetry, because gauge symmetry is the invariance under rotations in the abstract space of quantum fields.

Running on a track is a lot easier than running inside a swimming pool filled with water up to the level of your neck. The water resistance is much stronger than air resistance and your motion is slowed down. The Higgs substance can be viewed as a dense molasses-like fluid filling all space. As a particle moves inside this dense medium, its motion is affected. The particle experiences a resistance, which influences its inertia. This dragging felt by the particle inside the Higgs substance corresponds to the notion of *mass*. The analogy is not quite perfect, because the Higgs substance affects particles even when they are at rest, while fluid resistance does not. But, hopefully, this analogy will convey an image of the Higgs mechanism.

---

[22] T. Williams, *Cat on a Hot Tin Roof,* 1955.

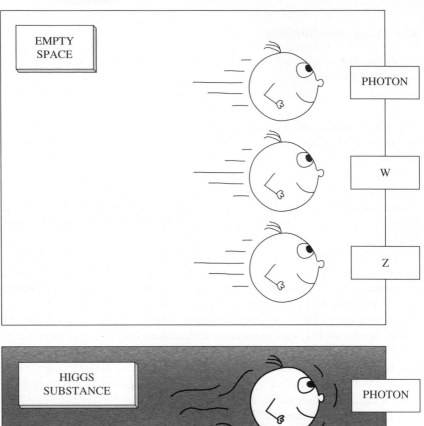

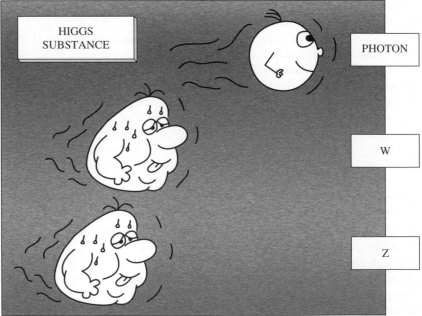

Figure 8.4 The photon, the *W*, and the *Z* would all propagate in the same way in empty space. But, when submerged in the Higgs substance, the *W* and *Z* become massive, while the photon remains unaffected.

If the Higgs substance did not exist, all elementary particles known to us would freely move in empty space and never be at rest, since their mass would be exactly zero. Mass arises as the response of particles living inside a space filled with Higgs substance. But not all particles react in the same way when they are submerged in the Higgs substance. As some people swim as swiftly as a dolphin while others are as clumsy as a hippopotamus, some particles are heavy because they interact strongly with the Higgs field, and some are light because they interact only weakly. The mass of an elementary particle is a measure of the strength of its interaction with the Higgs field. For instance, the W and Z behave like hippopotamuses when submerged inside the Higgs substance, while photons and gluons do not feel any effect at all. The Higgs substance is completely transparent for them, and the masses of photons and gluons remain zero even when they swim in this sticky medium.

Let us consider how the Higgs mechanism overcomes the problem of electroweak symmetry breaking, resolving the conflict between gauge symmetry and massive force carriers. When the W and Z move in space, they are hindered by the presence of the Higgs substance and they can propagate the weak force only up to a limited range of distances. In other words, they behave as massive force carriers. But gauge symmetry, although not manifest in the space filled with Higgs substance, is lurking inside the physical laws.

In the absence of the Higgs mechanism, the virtual starship of theoretical calculations stalled at the entrance of zeptospace. Lacking gauge symmetry, the engine of the virtual starship was not shielded from the dangerous effects of longitudinal waves. But the Higgs mechanism re-establishes gauge symmetry in the physical laws, even in the presence of massive W and Z. Effectively, the W and Z borrow parts of the Higgs field, camouflaging them as their longitudinal waves. Thanks to the Higgs mechanism the journey of the virtual starship can proceed through zeptospace and towards even smaller distances without encountering any difficulty. Gauge symmetry and massive W and Z are no longer incompatible, in the presence of the Higgs mechanism.

The Higgs mechanism not only explains the origin of particle masses, but it also sheds light on the meaning of electroweak unification. The difference that we perceive between weak and electromagnetic forces is just a consequence of the way in which the W, Z, and photon react to the Higgs substance. There is no conceptual difference between electromagnetic and weak forces, other than the accident that we happen to live in a space filled with a dense medium, in which the W and Z swim with great difficulty.

If the Higgs substance did not exist, the electroweak symmetry would be manifest in our world. Weak waves could be transmitted to very large distances and picked up by appropriate antennae, just as radios capture broadcast signals of electromagnetic waves. Instead, drowned in the Higgs substance, we are blind to the superb symmetry of forces in the particle world and we must use the power of our imagination and the strength of mathematics to infer the existence of a true unification of electroweak forces.

The Higgs mechanism may seem like a rather abstract idea, but it is completely analogous to a tangible phenomenon essential for the operation of the LHC: superconductivity. To understand the connection, let us consider a photon propagating inside a superconductor. Photons are associated with electromagnetic waves or, in other words, with oscillations of electric and magnetic fields. But, as seen in Chapter 6, superconductors have the property of expelling the magnetic field from their interior, a phenomenon called the Meissner effect. When photons propagate inside superconductors, the material reacts by trying to cancel the oscillating magnetic fields. As a result, photons are impeded in their motion and they can propagate the electromagnetic force only within a short range, behaving like massive force carriers. Photons inside a superconductor feel the same effect experienced by the W and Z particles inside the Higgs substance. In fact, the configuration inside a superconductor spontaneously breaks the gauge symmetry associated with electromagnetism, giving mass to the photon. In this respect the Higgs mechanism not only is one of the main goals of the LHC experiments, but it is even what makes the accelerator work, because superconductivity is the result of the spontaneous breaking of a gauge symmetry.

Indeed it was in the context of superconductivity that the condensed-matter physicist Philip Anderson (Nobel Prize 1977) first conjectured the existence of the Higgs mechanism. However, the real discovery of the Higgs mechanism in field theory was made in 1964 by Robert Brout and François Englert of the University of Brussels, and by Peter Higgs of the University of Edinburgh.

Robert Brout once told me that he had the idea while taking a shower. This may sound odd but, more often than not, theoretical physicists do not make their conceptual breakthroughs while they are sitting at a neat desk with a clean sheet of paper in front of them. Long periods of tenacious study and laborious calculations are necessary prerequisites, but then the right idea strikes at an unexpected moment, when the mind is more open to wandering through unbeaten paths.

About François Englert, Martin Veltman narrates: "At a dinner where Englert was also present I proposed a conjecture based on the statistics of one person (myself), namely that being born in the summer, preferably June, is the best with respect to intelligence. Englert, born in November, replied by saying that he was a Jew, and did not need this. Then he laughed so hard and I started to be worried for his life."[23]

---

[23] M. Veltman, *Facts and Mysteries in Elementary Particle Physics*, World Scientific, Singapore 2003.

Physicists have, in their own way, a lot of fun when they meet at conferences to discuss their theories.

Peter Higgs is a more secluded scientist, rarely seen at the conferences where theoretical physicists regularly gather. His historical publication has a peculiar genesis. During July 1964 he wrote two short notes on what we now call the Higgs mechanism. The second paper was rejected by the journal. Higgs sent the paper to a different journal and, in the revision process, added a short paragraph in which he pointed out the existence of new particles associated with the mechanism. That was the birth of the Higgs boson. Forty-five years later we are still looking for experimental confirmation that the Higgs mechanism is responsible for electroweak symmetry breaking.

Experimental search

> It is a terrible thing for a man to find out suddenly that all his life he has been speaking nothing but the truth.
>
> Oscar Wilde[24]

How can the LHC test the reality of the Higgs mechanism? Since the Higgs mechanism is based on the existence of a new substance filling all space, the most convincing proof would be to remove this substance, say from a small region of space, and then check what happens to matter in that region. As explained later, mass would not disappear with the removal of the Higgs substance, because the Higgs mechanism contributes less than a kilogram to the mass of an average person.

At this point you might think that the removal of the Higgs substance is an easy way of effortlessly losing a bit of extra weight, but removing the Higgs substance would be highly ill-advised from the point of view of health concerns. The mass difference between neutrons and protons critically depends on the Higgs substance. Even a small change in the density of the Higgs substance would result in a modification of the mass difference between neutrons and protons with dramatic consequences for our world, since no stable molecules could exist, no chemical or nuclear processes could be sustained, and all matter would collapse into simple sterile atoms. We could not live without the Higgs substance.

At any rate, modifying the density of the Higgs substance, aside from having catastrophic effects, is practically impossible. It could be done only by heating the universe to temperatures above 10^{15} degrees, a value one hundred million times larger than the temperature in the centre of the sun. It is very unlikely that these enormous temperatures will be

---

[24] O. Wilde, *The Importance of Being Earnest*, 1895.

ever attained during the existence of humanity, even under the most pessimistic extrapolations of global warming. The Higgs substance is here to stay and no human being will ever be able to modify it. It is necessary to devise an alternative strategy to experimentally test for the Higgs mechanism.

Every quantum field can develop ripples, which we call particles, by localizing energy in a small region of space. The Higgs field is no exception. While the Higgs substance corresponds to a uniform distribution of the Higgs field in space, it is possible to disturb the field and create small ripples over the calm sea of Higgs substance. These ripples correspond to a new kind of particle called the *Higgs boson*.

Although no Higgs boson has yet been observed, the theory is able to predict all its properties, except for one parameter: its mass. Since the mass of an elementary particle is a measure of its interaction with the Higgs field, it is not surprising that the Higgs boson mass corresponds to the self-interaction of the Higgs field. Experiments prior to the LHC have searched for the Higgs boson, but have failed to discover it. Their unsuccessful searches have been used to exclude certain ranges of mass. Experiments at LEP have ruled out the existence of a Higgs boson with mass less than 114 GeV. In 2009, experiments at the Tevatron reported the exclusion of a Higgs boson with mass in the range between 160 and 170 GeV. Moreover, theoretical arguments based on the consistency between Standard Model predictions and LEP measurements disfavour a Higgs boson with mass larger than about 200 GeV.

The Higgs boson, if it really exists, can be produced and detected at the LHC, whatever its mass might be. But experiments cannot directly observe the Higgs boson because it is a highly unstable particle. In less than 10^{-22} seconds it transforms its energy into other kinds of particles or, as physicists say, it decays. The identification of the Higgs boson at the LHC can only be made through the detection of the particles produced in its decay: hadronic jets, leptons, and photons. The problem is that of course there are many other sources of jets, leptons, and photons in collisions at the LHC, and the difficulty lies in distinguishing particles produced in Higgs decays from identical particles produced in other conventional processes. Physicists call *background* all the conventional processes that can potentially mask the *signal*, which are the events generated by the genuine creation of a new particle, in this case the Higgs boson.

Physicists have to analyse a very large number of collisions to select the events that are potentially interesting. The experimental data are then compared with the output from elaborate computer programs that simulate high-energy proton collisions and the response of the LHC detectors, predicting both the background and the signal. These simulation programmes are the result of complex theoretical calculations and of years of testing the instrumentation used in particle detectors. They

Figure 8.5 The simulation of an LHC event with production of the Higgs boson in the CMS detector.

Source: CERN/CMS Collaboration.

are very sophisticated tools, whose methodology has been proven effective and reliable in the interpretation of data from previous experiments at LEP, HERA, and the Tevatron. But before a new discovery is claimed, many cross checks must be completed with analyses that go beyond a simple comparison between data and simulations.

Experimentalists perform various procedures of internal validation of their results. Once the response of the different instruments in the detectors is fully tested and understood, various techniques are employed to extract the background evaluation mostly from data, relying as little as possible on numerical simulations. Only after a long and delicate process of studying the detector performance and of analysing the data, can experimental physicists be sure that a signal is observed over the background and, after years of hard work, can finally announce to the world the discovery of the Higgs boson.

How to understand nothing

To think is difficult. To think about nothing is more difficult than about something.

Lev Okun[25]

[25] L.B. Okun, *Vacua, Vacuum: The Physics of Nothing, in History of Original Ideas and Basic Discoveries in Particle Physics,* ed. H.B. Newman and T. Ypsilantis, Plenum Press, New York 1996.

The goal of collider experiments is not just the discovery of some new particles, but it is the identification of the principles that can guide us towards an understanding of nature and its fundamental laws. During the 1950s and 1960s there was a deluge of new hadrons discovered, but little progress in the understanding of their significance. At that time Willis Lamb declared: "The finder of a new elementary particle used to be rewarded by a Nobel Prize, but such a discovery now ought to be punished by a $10,000 fine."[26] But the deeper we penetrate into the essence of matter, the more we find that new principles are associated with new particles. Therefore, in some cases, the detection of a new particle really entails a fundamental discovery. This will certainly be the case if the Higgs boson is found.

The discovery of the Higgs boson would validate our ideas about spontaneous symmetry breaking or, in other words, on the nature of the vacuum, a concept that has been debated since antiquity. Aristotle, in contrast with the atomists, claimed that empty space could not exist in nature: "There is no void existing separately, as some maintain."[27] He even offered a proof of this statement, starting from the "self-evident" assertion, based on empirical observation, that any body in motion comes to a stop unless some external force is applied to it. But, argued Aristotle, in empty space all points are equal and a body could not decide where to stop; then motion would last forever, which is absurd. Hence, concludes Aristotle, empty space cannot exist. In his own words: "Nobody can give a reason why a body that has been put into motion in empty space should stop on its own account. Why should it stop in one place rather than in another? Thus it will either remain at rest, or it will of necessity keep moving ad infinitum unless it is hindered from doing so."[28] Moreover, continues Aristotle, in the absence of any external agent, bodies follow their natural motion, which makes heavy elements (earth and water) fall down and light elements (air and fire) go up. How could natural motion exist in empty space?

It is interesting to note that, had Aristotle turned the argument around and assumed the existence of empty space, he would have discovered the principle of inertia two thousand years ahead of time. Instead, Aristotle's conclusion was elevated to a principle, later called *horror vacui* ("dread of emptiness", in Latin). Although the name sounds like some obscure form of morbid phobia, the principle actually states that the forces of nature act in such a way as to prevent the creation of vacuum.

Before Aristotle, in the fifth century BC, Empedocles offered an experimental proof of horror vacui, which is sometimes referred to with

---

[26] W.E. Lamb, *Nobel Lecture Speech,* 12 December 1955.
[27] Aristotle, *Physics*, Book IV.   [28] Aristotle, *ibid.*

the poetic name of Empedocles' Hydra. Take a vessel with two holes, one at the top and one at the bottom, and fill it with water. As long as you keep the top hole closed, no water flows from the bottom hole. Why? Horror vacui is stronger than gravity and keeps the water up, or else vacuum would be created inside the vessel. But as soon as you open the top hole, water freely flows down because there is no danger of creating any vacuum.

Figure 8.6 Torricelli's experiment disproving *horror vacui*: the columns of mercury fall at the same height in all containers, independently of the volume left empty.

Source: Photographic Department, Institute and Museum of the History of Science, Florence / Photo Eurofoto.

As explicitly stated in the *Dialogue*, Galileo believed that vacuum could exist, at least ideally as the limit in which a medium becomes increasingly diluted. The actual physical reality of vacuum – which would have contradicted Scholastic tradition – is not discussed. Nonetheless, Galileo also believed in horror vacui, though not as an absolute principle, but as a force that can be quantified, and he performed many interesting experiments measuring its effects.

In 1644, only two years after Galileo's death, one of his pupils, Evangelista Torricelli, made the decisive experiment proving that it is possible to evacuate certain regions of space. He filled some long and thin containers up to the rim with mercury, a liquid much heavier than water. Then he turned them upside down, submerging their open ends into a vessel also filled with mercury. He observed that mercury was falling to exactly the same height in all containers, independently of the shape and size of the volume left empty. This was the proof that horror vacui cannot hold, because it would exert in the various containers different forces, in proportion to the amount of the space left empty.

Later experiments, notably by Blaise Pascal, demonstrated conclusively that most of the phenomena imputed to horror vacui are actually due to air pressure. That was the end of horror vacui: nature doesn't abhor emptiness. To most of us today the concept of vacuum seems fairly intuitive and not at all mysterious. Although we cannot see air with our eyes, our intuition is accustomed to the effect of air pressure. Take away everything and you are left with nothing: that's vacuum.

But modern science has shown that things are not so simple. Take a cavity and remove all the matter from its interior, and you are not left with "nothing". The walls of the cavity emit an electromagnetic radiation that permeates through the interior of the cavity. This radiation depends on the temperature of the walls and cannot be sucked out by any pump. It is called *black-body radiation*. Our universe, even in the most isolated and empty regions, is filled with a nearly uniform blackbody radiation, called *cosmic microwave background*, that has cooled since the time of the Big Bang and has now reached the temperature of −270°C.

Even if we imagine cooling a region of space to absolute zero temperature, thereby eliminating the black-body radiation, we are not left with "nothing". Quantum mechanics, through a phenomenon that will be described in Chapter 9, produces continuous field fluctuations creating particles for very short periods of time. The vacuum in quantum mechanics is a very busy place, quite different from empty space. The quantum fields are never perfectly still but, like the real sea, constantly fluctuate generating small ripples that quickly come and go.

If the LHC discovers the Higgs boson, the complexity of the vacuum will acquire a new element. In physics, vacuum is not empty nothingness: vacuum is the configuration of minimal energy of a system. The discovery of the Higgs boson will prove that nature has chosen a vacuum not made of "nothing" but of "something", because energy is saved this way. This "something" is an entity uniformly filling all space: the Higgs substance. The LHC is trying to make this substance vibrate slightly, to create a few ripples on it and to detect them in the form of a Higgs boson. By finding this new facet of the physical vacuum, the LHC will discover that nature abhors nothingness and desires to fill emptiness. Paradoxically, the LHC will demonstrate that the basic idea of horror vacui was right, after all.

Open questions

> He who asks a question is a fool for five minutes; he who does not ask a question remains a fool forever.
>
> Chinese proverb

The most inappropriate name ever given to the Higgs boson is "the God particle". The name gives the impression that the Higgs boson is the central particle of the Standard Model, governing its structure. But this is very far from the truth. And yet this name, although never used in physics terminology, seems very popular among newspaper journalists.

Sheldon Glashow gave a much more appropriate definition of the Higgs boson: "Sometimes I compare today's very successful theory of elementary-particle physics with a gorgeous and elegantly crafted mansion. But every residence, humble or grand, must contain an object of no great beauty... The flush toilet is a rather ugly thing, but it works and no one has come up with a plausible alternative."[29] The Higgs boson is like the toilet of the Standard Model edifice. Although indispensable for the functioning of the house, it isn't something that you proudly show to your guests.

Our present understanding of matter and forces is based on three elements: general relativity (which describes gravity), Yang–Mills gauge theory (which describes strong and electroweak forces and the composition of matter), and the Higgs sector (which describes the spontaneous breaking of electroweak symmetry). The elegance and simplicity of general relativity and Yang–Mills gauge theory is undisputable. Both of these theories are fully dictated by the logical consequence of gauge symmetry, have very few adjustable parameters, and fare marvellously when compared with experimental data. In spite of our unsuccessful attempts to convincingly unify these two elements into a single theory, there is little doubt that general relativity and gauge theory belong to one of the highest steps of Jacob's ladder.

---

[29] S. Glashow, *Interactions,* Warner Books, New York 1988.

The *Higgs sector* is that part of the theory that describes the Higgs mechanism and contains the Higgs boson. Unlike the rest of the theory, the Higgs sector is rather arbitrary, and its form is not dictated by any deep fundamental principle. For this reason its structure looks frightfully ad hoc. The conventional implementation of the Higgs sector into the Standard Model corresponds to the simplest possible choice one can make for its structure. This minimal choice leads to the supposition of a single Higgs boson. But nothing forbids other more elaborate schemes, which predict the existence of various types of Higgs bosons or even new kinds of phenomena.

The Higgs sector explains the structure of quark and lepton masses that we observe, but only at the price of introducing 13 adjustable input parameters determined by experimental measurements. Quark and lepton masses can certainly be accounted for by the Higgs sector, but unfortunately the theory is unable to predict their values. Moreover, although the Higgs sector can generate the spontaneous breaking of electroweak symmetry, it provides no deep explanation about the force that is ultimately responsible for the phenomenon.

The Higgs sector, in contrast with general relativity and gauge theory, has not yet been experimentally confirmed. This gives us hope that all the unsatisfactory aspects of the Higgs sector are only the consequence of our limited knowledge and not of a poor choice made by nature. For this reason the experimental investigation of the Higgs boson at the LHC is crucial for making progress in our understanding of the particle world. The Higgs boson may turn out to be quite different from our expectations. New experimental information is urgently needed to find the right track that can lead us to a deeper explanation of the phenomenon of electroweak symmetry breaking.

It is sometimes said that the discovery of the Higgs boson will explain the mystery of the origin of mass. This statement requires a good deal of qualification. Most of the mass of ordinary matter is carried by atomic nuclei, which are made of protons and neutrons which, in turn, are made of quarks. But the masses of protons and neutrons are not simply given by the sum of the masses of the constituent quarks, which accounts for only about 1 per cent of the total. Mass is (recall $E = mc^2$) the intrinsic energy of a body at rest. So about 98 per cent of the mass of protons or neutrons comes from the frantic motion of quarks and gluons confined in their interiors or, more precisely, from the binding force of QCD. Electromagnetic effects count for about another 1 per cent.

The Higgs mechanism is ultimately responsible for generating the quark masses, but not for the QCD effect. This is the reason for which it was stated previously that the Higgs substance provides for less than a kilogram of our body mass. Moreover, as will be illustrated in Chapter 12, most of the matter in the universe is in the form of dark matter. Although the nature of dark matter is still unknown, it is unlikely that its mass originates from the Higgs substance. In summary, the Higgs mechanism accounts for about 1 per cent of the mass of ordinary matter, and for only 0.2 per cent of the mass in the universe. This is not nearly enough to justify the claim of explaining the origin of mass.

On the other hand, the Higgs mechanism generates the masses of all known elementary particles and, in this regard, is a crucial source of mass in the particle world. However, the real mystery in the origin of elementary particle masses lies in the structure of quark and lepton masses. This structure follows a very distinctive pattern, which cries out for an explanation. Unfortunately, after decades of research, theoretical physicists have not yet cracked the problem and have made almost no progress in finding a theory in which the observed pattern of masses can be deduced from basic principles. It is very unlikely that the discovery of the Higgs boson alone could be the key that unlocks the mystery of quark and lepton masses.

The importance of the Higgs mechanism lies in the origin of a fundamental length scale, which is referred to as the *weak length*. The weak length, set by the density of the Higgs substance, defines the range of the weak force, equal to about 10^{-18} m. Fundamental lengths play a crucial role in physics because they determine the locations of the steps in Jacob's ladder. In other words, they determine the transitions between different theoretical descriptions of nature. They are the gateways toward a deeper understanding of nature.

There is little doubt that the discovery of the Higgs boson will be a fundamental step in our understanding of the particle world, although it will elucidate the problem of symmetry breaking more than that of mass. But the Higgs boson leaves too many unanswered questions to believe that it represents the final answer. All the shortcomings that afflict the Higgs sector indicate that the discovery of the Higgs boson is more likely to be a starting point for new explorations, rather than the ultimate landing place of knowledge.

9
Dealing with Naturalness

Life is full of infinite absurdities, which, strangely enough, do not even need to appear plausible, since they are true.

Luigi Pirandello[1]

When the lunar module landed on the surface of the moon on 20 July 1969, Neil Armstrong and Buzz Aldrin didn't expect to be welcomed by strange creatures with green skin and long antennae. Enough was already known about the moon to exclude such a singular encounter. Rather, the exploration of zeptospace undertaken by the LHC can be likened to the voyage made by Marco Polo. The Venetian traveller knew that Cathay existed, but had only a vague idea of the "fabulous cities and strange beasts" he was hoping to meet on his way. Similarly, we know that something new must exist in zeptospace: the element responsible for electroweak symmetry breaking. Most likely this new element will take the form of a Higgs boson. But many physicists are convinced that that the Higgs boson cannot be the end of the story and that zeptospace must be populated by other "fabulous particles and strange phenomena".

This chapter explains the main reason why new particles, beyond the Higgs boson, are believed to lurk in zeptospace, while the following two chapters describe some of the ideas of what physicists imagine zeptospace looks like. While up to here I have been mainly presenting facts in physics, now I cross the border between reality and speculation, and enter the territory of pure theoretical hypotheses. The theories we will encounter, although based on reasonable physics principles and on the sparse experimental information about zeptospace gathered from previous colliders, are still only conjectures, which represent attempts to extrapolate our present knowledge to the uncharted land of zeptospace. The ultimate arbiter – the LHC – will rule out most (if not all) of these ideas. But perhaps the LHC will confirm one of them, demonstrating, once again in the history of science, how powerful human

---

[1] L. Pirandello, *Six Characters in Search of an Author,* 1921.

imagination is. However, it could well be that nature is much more creative than our minds and that zeptospace is designed in a way that our
imagination could not grasp. This outcome would only make the voyage
of the LHC into zeptospace an even more fascinating adventure.

The hierarchy

> I am ill at these numbers.
> William Shakespeare[2]

While experimental physicists are busy driving their starship – the
LHC – through the border between the known world and zeptospace,
theoretical physicists are surreptitiously running ahead, dreaming about
realities at much smaller distances. Suppose that the Higgs is discovered
at the LHC. Then the Standard Model can be validly extrapolated way
beyond zeptospace, towards very small distances. But how small?

Nothing forbids theorists to imagine colliders much more powerful
than the LHC. In their minds they can accelerate protons in rings larger
than the earth, bending beams with fantastic magnetic fields and
producing collisions of stupendous energies. These prodigious colliders
are like the virtual starship that can travel into unimaginably small
distances. But, at a certain point, even the wildest imagination hits a
brick wall and cannot go any further. Once we reach distances as small
as 10^{-35} m, the so-called *Planck length*, the engine of our virtual starship
breaks down completely. At less than the Planck length the theory of
particle physics is no longer able to describe any physical entity. As a
matter of fact, if our ideal collider is so powerful as to smash particles
and pack them within the Planck length, their gravitational attraction
becomes so strong that the system collapses into a black hole. The black
hole gobbles up all information and we have no means of knowing what
is going on at distances smaller than the Planck length. At these distances
our concepts of space and time break down. Gravity becomes so intense
that the Standard Model must be replaced by a new coherent description
of general relativity and quantum mechanics, of gravity and gauge
forces.

The Planck length is incredibly small. It is small even in comparison
with the tiny distances familiar to particle physics. The smallest distance
directly explored so far by experiment is roughly equal to the *weak
length*, the range of the weak force, which is about 10^{-18} m. The Planck
length is 10^{17} times smaller than the weak length, which means that it
compares with the weak length as the size of a human compares with

---

[2] W. Shakespeare, *Hamlet*, Act II, Scene II.

the distance from here to Sirius, the star in the Canis Major constellation. The largeness of the ratio between the weak and the Planck lengths (equal to 10^{17}) is usually referred to as the *hierarchy between weak and gravitational forces.*

For physical processes involving particles of relatively low energy, the weak and Planck lengths measure the strengths of weak and gravitational forces, respectively. Therefore, the hierarchy expresses the fact that, in particle-physics experiments, gravity is so feeble that it is absolutely negligible with respect to the weak force, and consequently also with respect to the electromagnetic and strong forces. In the particle world, gravity is the weakest of all forces. The astronomically large disparity between gravity and the other forces brings in a question: is there a fundamental reason why there is such an enormous hierarchy between weak and gravitational forces?

This is one of those questions that aims at investigating not *how* things work, but *why* things work in the way we observe them. The reason we address this question is based on the belief that any physical constant will eventually find its final explanation in the context of a truly unified theory, in which all parameters can be computed. The largeness of the number describing the hierarchy between weak and gravitational forces is so striking that it looks like a clue for some concealed aspect of the final theory. But the most puzzling aspect about the largeness of the hierarchy surfaces only once we consider the odd world of quantum mechanics.

A quantum puzzle

> Anyone who is not shocked by quantum theory has not understood a single word.
>
> Niels Bohr[3]

Heisenberg's principle states that there is always a trade-off on how two complementary physical quantities can be determined. The more precisely one of the two quantities is known, the more uncertain the other quantity becomes. For instance, when we measure the energy of a particle within a specific time interval, we cannot determine with absolute precision *both* energy and time. Thus, in practice, there always exists an intrinsic uncertainty in the determination of the energy of a system, independently of the quality of the instruments used in the measurement. The physical reality of a phenomenon cannot, even ideally, be known with absolute certainty and precision.

---

[3] N. Bohr, as quoted in M.J. Wheatley, *Leadership and the New Science: Discovering Order in a Chaotic World*, Berrett-Koehler, San Francisco 1999.

Figure 9.1 Werner Heisenberg (centre) with Wolfgang Pauli (left) and Enrico Fermi on a boat on Lake Como in 1927.

Source: Pauli Archive / CERN.

Heisenberg's uncertainty principle, which incarnates the essence of the indeterministic character of quantum mechanics, is not a speculative idea but a well-established property of nature and explains some apparently paradoxical phenomena in particle physics. One of the best examples is alpha radioactivity. Radioactive nuclei emit alpha particles (helium nuclei) at relatively low energies. For instance, the radioactive emission in the uranium isotope U^{238} has an energy of 4.2 MeV. And yet, if much more energetic alpha particles are shot towards uranium targets, they cannot penetrate inside the nuclei and they are repelled by the repulsive electromagnetic force that acts between charges of the same sign. How is it possible that the alpha particles emitted by radioactivity are not accelerated to energies much greater than 4.2 MeV by this repulsive force? This is as disconcerting as a roof tile falling onto your head from a five-storey building with the gentle delicacy of an autumn leaf. Why doesn't the tile accelerate through gravity and break your skull?

This paradoxical situation in the context of classical physics is perfectly legitimate in quantum mechanics. Because of Heisenberg's principle, the energy of the alpha particle can fluctuate even to very large values, as long as this fluctuation lasts for a sufficiently short time. The alpha particle borrows the energy that allows its escape from the nucleus and then hands it back very quickly, just in time to comply with Heisenberg's principle.

Unscrupulous brokers sometimes act in the same way as alpha radio-activity by "short selling". The broker sells a financial share that he actually does not own, with the intent of later purchasing it at a lower price. The operation must be done fast enough for the lender not to realize the momentary deficit. In physics, this process is called the *tunnel effect*, because it visually corresponds to passing through a mountain without having the energy to climb it. Though contrary to ordinary intuition, the tunnel effect is a familiar phenomenon in quantum mechanics. It is used, for instance, in fast electronic components such as the semiconductor diodes invented by Leo Esaki (Nobel Prize 1973). These diodes switch currents on and off so rapidly that they can be used to build oscillators operating at frequencies above 100 GHz.

Heisenberg's principle is also at the origin of bizarre phenomena related to the existence of *virtual particles*. A virtual particle has exactly the same intrinsic properties of an ordinary particle (same mass, same electric charge, and so on) but has a completely abnormal value of energy. The energy of a virtual particle is plain "wrong". Mass and velocity unambiguously determine the "right" energy of a particle. The faster a particle moves, the larger its energy. But this is no longer true for a virtual particle: its energy is completely independent of its velocity and can take any value. For instance, a virtual electron can carry a colossal amount of energy, and yet move very slowly. Even the most acclaimed Brazilian soccer player would risk a total fiasco in a game played with a "virtual" ball. Despite the large amount of energy imparted by his powerful kick, there is a chance that the ball would hardly move, as if kicked by a toddler who has just learned how to walk.

The life of a virtual particle lasts only for an extremely short time. According to Heisenberg's principle, the larger the energy of a virtual particle, the shorter is its existence. This makes virtual particles practically invisible to our perception, thereby saving the reputation of Brazilian soccer players.

Virtual particles are the source of many extraordinary phenomena in the world of particle physics. In particular, their presence makes empty space a very busy place. Pairs formed by virtual particles and antiparticles come to existence in empty space and then vanish fast enough to comply with Heisenberg's principle. The quantum vacuum – the empty space of quantum mechanics – is populated by their constant appearance and disappearance, because energy conservation does not preclude their short existence. Just as ephemeral spirits haunt the empty halls of old Scottish castles, so virtual particle–antiparticle pairs pop out of nowhere and swiftly vanish in the queer house of the quantum vacuum.

The complexity of the quantum vacuum reveals a new and much more dramatic facet of the hierarchy between weak and gravitational forces: the *naturalness problem*. Most of the dreams about the "fabulous cities and strange beasts" that may exist in zeptospace have been prompted by attempts to resolve this problem.

The naturalness problem

> Everything is natural: if it weren't, it wouldn't be.
>
> Mary Catherine Bateson[4]

According to the Higgs mechanism, space is filled with a uniform density of Higgs substance. But because of the complexity of the quantum vacuum, the calm sea of Higgs substance is continually disturbed by the rapid production and annihilation of all sorts of virtual particles. The constant buzz of virtual particles affects the density of the space-filling Higgs substance. Just as ghosts flickering in and out of the netherworld leave vivid memories in the minds of impressionable individuals, so virtual particles leave their indelible imprint in the Higgs substance. The swirling of virtual particles effectively gives a gigantic contribution to the density of the Higgs substance, which becomes extremely thick.

Theoretical calculations show that this contribution to the density of the Higgs substance is proportional to the maximum energy carried by virtual particles. Since virtual particles can carry huge amounts of energy, the molasses-like Higgs substance becomes thicker than mud or even as hard as rock when quantum-mechanical effects are taken into account. Ordinary particles moving inside this medium should feel a tremendous resistance or, in more precise physical terms, should acquire enormous masses. Calculations based on a simple extrapolation of the Standard Model all the way down to the Planck length yield the result that electrons should be one million billion times more massive than what we observe – as heavy as prokaryote bacteria. But since electrons are obviously not as heavy as bacteria, we are confronted with a puzzle: why is the Higgs substance so dilute in spite of the natural tendency of virtual particles to make it grow thicker?

This dilemma is usually referred to as the *naturalness problem*. The density of the Higgs substance determines how far the W and Z particles can propagate the weak force; in other words, it determines the weak length. Therefore, when virtual particles make the Higgs substance thicker, they effectively make the weak length shorter. The naturalness problem then refers to the conflict between, on one side, the tendency

---

[4] M.C. Bateson, *On the Naturalness of Things*, in *How Things Are: A Science Toolkit for the Mind*, ed. J. Brockman and K. Matson, William Morrow & Co, New York 1995.

of virtual particles to make the weak length as small as the Planck length and, on the other side, the observation that the two length scales differ by the enormous factor of 10^{17}. Let us rephrase the problem with an analogy.

Suppose that you insert a piece of ice into a hot oven. After waiting for a while, you open the oven and you discover that the ice is perfectly solid and hasn't melted at all. Isn't it puzzling? The air molecules inside the hot oven should have conveyed their thermal energy to the piece of ice, quickly raising its temperature and melting it. But they did not.

The naturalness problem is equally puzzling. The energetic virtual particles are like the hot air molecules of the oven analogy, and the Higgs substance is like the piece of ice. The frenzied motion of virtual particles is communicated to the Higgs substance, which should become as hard as rock. And yet, it remains very dilute. The weak length should become as small as the Planck length. And yet, the two lengths differ by a factor of 10^{17}.

Just as in the inside of a hot oven nothing can remain much cooler than the ambient temperature, so in the quantum vacuum virtual particles do not tolerate that the weak length remain much larger than the Planck length. Thus, the real puzzle is that no hierarchy between weak and gravitational force should exist at all, let alone there being a difference by a factor of 10^{17}. The essence of the naturalness problem is that the anarchic behaviour of virtual particles does not tolerate hierarchies.

At this point, a very important warning should be issued. The naturalness problem is not a question of logical consistency. As the word says, it is only a problem of *naturalness*. Virtual particles provide one part of the energy stored in the Higgs substance. Nothing forbids the possibility of nature carefully choosing the initial density of the Higgs substance in such a way as to nearly compensate the effect from virtual particles. Under these circumstances, the enormous disparity between the weak and Planck lengths could be just the result of a precise compensation among various effects. Although this possibility cannot be logically excluded, it seems very contrived. Most physicists have difficulties accepting such accurate compensations between unrelated effects, and regard them as extremely *unnatural*.

It may appear surprising that something so vague and subjective as the concept of naturalness can find a place in rigorous physics theories that employ the most sophisticated mathematics. But theoretical physicists often follow a sense of aesthetic beauty, purity, and simplicity in formulating their creations, just as in art or in other speculative human activities. The odd thing is not that theoretical physicists use beauty and simplicity as sources of inspiration, but rather that nature seems to follow the same principles too. When Einstein was asked what he would have done, had Eddington's observation of the 1919 solar eclipse disproved, rather than confirmed, his theory, he simply replied: "Then

I would have felt sorry for the dear Lord."[5] Clearly he was confident that general relativity was too beautiful for nature to shy away from.

Aesthetic beauty and naturalness are powerful inspirational principles but, of course, they cannot be used to validate a theory. Moreover, since they are subjected to philosophical influences, sometimes they can even be misleading. Take the example of the Copernican and Ptolemaic systems. Even leaving aside any empirical evidence, modern scientists find that the solar system is more *naturally* explained by a heliocentric theory, in which simple elliptic orbits describe planetary motions, rather than by a geocentric theory, which requires the introduction of different epicycles for each planet. But to predecessors and contemporaries of Copernicus a geocentric theory probably appeared more *natural*. Tycho Brahe discarded a heliocentric description of the solar system with the subjective argument that the earth is a "hulking, lazy body, unfit for motion."[6] History of science abounds with pitfalls of false prejudices.

But the naturalness problem regarding the density of the Higgs substance has some peculiar features that make most physicists believe that it is not based upon a false prejudice and that there must be some deep truth in it. Certainly we cannot logically exclude a fortuitous compensation of the various contributions able to give a very dilute Higgs substance. But this compensation would require a stupendous coincidence or *fine-tuning*, as is usually said in physics parlance.

One talks of fine-tuning when different parameters of the theory conspire to give very delicate and precise cancellations capable of making the total effect much smaller than individual contributions, as a result of a mere fortuitous coincidence. The naturalness problem boils down to an issue of fine-tuning. Maintaining the large hierarchy between the weak and Planck lengths in the presence of the quantum vacuum is not logically impossible, but it requires an unexplained fine-tuning of the parameters of the Standard Model with an accuracy of one part in 10^{34}. To get a feeling of how ridiculously contrived such a fine-tuning appears, let me offer an example.

Suppose you enter a room and find that somebody left on the desk a pencil standing upright on its tip. This highly unstable position of the pencil looks very unnatural. Somebody must have fine-tuned its position in such a way that the pencil's centre of mass falls exactly within its small tip. The longer the pencil, the higher is the degree of fine-tuning

---

[5] A. Einstein, as quoted in I. Rosenthal-Schneider, *Reality and Scientific Truth: Discussions with Einstein, von Laue, and Planck*, Wayne State University Press, Detroit 1980.

[6] *Tychonis Brahe Dani Opera Omnia*, ed. J.L.E. Dreyer, Hauniæ: In Libraria Gyldendaliana, Copenhagen 1913–1929.

required to maintaining the pencil in this awkward position. A fine-tuning of one part in 10^{34} corresponds to poising a pencil as long as the solar system on a tip 0.1 mm wide!

It just seems implausible that the existence of our universe rests on such an extraordinary coincidence. Most particle physicists hold the belief that behind each apparently mysterious coincidence and each incredibly accurate fine-tuning must lie some good explanation. The hierarchy between weak and gravitational forces cannot be just a fortuitous accident, but it must hide some profound significance. There must be an invisible hand that keeps the pencil upright, converting a seemingly inconceivable coincidence into a perfectly logical result.

There is another reason why most physicists believe that the naturalness problem bears an important significance. Virtual particles can carry any amount of energy. Therefore a fortuitous compensation of the various contributions to the density of the Higgs substance would require special correlations between phenomena occurring at widely different scales of energy. Metaphorically speaking, all steps of Jacob's ladder should be correlated with excruciating precision. Although not logically excluded, this possibility goes against our intuition that each step of Jacob's ladder can be treated separately and challenges our basic ideas of effective field theories.

The essence of the naturalness problem lies in the unruly behaviour of virtual particles. Their ideology is summarized by an old principle of physics: "Everything which is not forbidden is compulsory." Like restless children, virtual particles do every possible mischief that is not strictly prohibited. The mischief we are concerned with is making the Higgs substance absurdly thick and, consequently, the weak length as small as the Planck length.

Most physicists believe that the naturalness problem is an indication that the Standard Model is incomplete. A new theory should replace the Standard Model as we enter zeptospace, a theory able to modify the behaviour of virtual particles in such a way that they no longer mess up the Higgs substance. What could bring virtual particles to behave in a more disciplined and obliging way?

There is actually one law that the anarchic virtual particles are ready to obey, and that is symmetry. Virtual particles show respect only for physical quantities that participate in symmetry transformations, but happily disregard the rest. Unfortunately the parameter that determines the density of the Higgs substance does not participate in any symmetry and this is the root of the naturalness problem.

Symmetry could be the invisible hand that solves the dilemma. The idea that some new symmetry or some new theoretical element could resolve the naturalness problem has proved to be one of the most fruitful sources of inspiration for theoretical physics in the past decades, and has led to many imaginative and original proposals.

10
Supersymmetry

If I seem unduly clear to you, you must have misunderstood what I said.

Alan Greenspan[1]

Space-time is the arena in which natural phenomena occur. But with a leap of imagination let us conceive a vaster space that extends in dimensions beyond the ordinary three directions of space and the one of time. Moreover, let us imagine that the geometry of the space described by these new dimensions is completely unconventional. Every square, no matter how long its sides are, has an area equal to zero. Rectangles can have non-zero areas, but have unfamiliar properties too. As explained in Figure 10.1, if the sides of a rectangle are switched, its area changes sign and becomes negative. Geometry in this space is more perplexing than in an Escher drawing or in a Dalí painting. It is quite impossible to visualize this space with simple pictures on a piece of paper, because its dimensions have a true "quantum-mechanical" nature. But the power of mathematics goes beyond the grasp of our visual perception and allows us the exploration of this surrealistic space, which is called *superspace*.

Although as odd as one can think of, superspace is a logically consistent entity and we can indulge ourselves with dreams about what matter would look like in superspace. Particles in superspace are so different from particles in ordinary space that they deserve their own special name: *superparticles*. A superparticle is an odd entity. We can picture it as Janus, the Roman god of beginnings and ends, who had two faces looking in opposite directions. Like Janus, each superparticle has a double identity being at the same time two particles with different spin.

Spin is an intrinsic rotational motion of particles. In 1925 two Dutch students, Samuel Goudsmit (1902–1978) and George Uhlenbeck (1900–1988), invented the concept of particle spin, imagining electrons

[1] A. Greenspan, address to a Senate Committee in 1987, as quoted in *Guardian Weekly*, 4 November 2005.

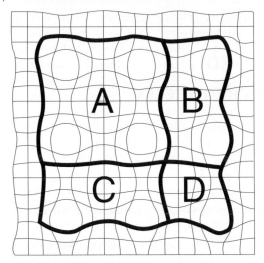

Figure 10.1 Squares have zero areas in superspace. Thus, area(A) = 0 and area(D) = 0, but also area(A + B + C + D) = 0. Hence, one deduces that area(B) = −area(C). The area of a rectangle changes sign when its sides are switched, and negative areas are possible. The figure is only meant as an illustration, because superspace cannot be simply visualized.

as tiny spinning tops. However, as Hendrik Lorentz quickly pointed out to them, the idea of spinning electrons runs into many logical paradoxes and it is absolutely untenable. For instance, Lorentz explained to the two young physicists that the edges of an electron spinning in the way envisaged by them would rotate at a speed faster than light, in blatant contradiction with the principle of relativity. Goudsmit and Uhlenbeck were so taken aback by the arguments of the old master of Dutch physics that they immediately asked their advisor, Paul Ehrenfest, to withdraw their article. But Ehrenfest told them that it was too late because he had already sent the article to the journal, adding: "You are both young enough to be able to afford a stupidity."[2]

But those were the times of quantum mechanics when a certain amount of recklessness didn't hurt for making discoveries. Although the arguments by Lorentz made perfect sense in terms of classical physics, they were not valid in quantum mechanics, where phenomena defeat our intuition. Particle spin is a physical reality, although it is a concept that cannot be expressed in terms of classical physics. Spin is like an incessant intrinsic rotational motion of particles, but it eludes a simple classical description. The rate at which a particle is spinning never changes and thus spin is an individual characteristic of a particle, like its electric charge or its mass. Moreover, as for other physical quantities in quantum mechanics, spin can exist only in integer multiples of a fundamental amount (called spin ½). Particles with spin equal to an odd multiple of this amount are called *fermions* – from the name of Enrico Fermi, who studied their statistical properties. Particles with zero spin

---

[2] G.E. Uhlenbeck, *Physics Today* 29, 43 (1976).

or an even multiple of the fundamental amount are called *bosons*, from the name of the Indian physicist Satyendra Nath Bose (1894–1974). Quarks and leptons carry spin ½ and belong to the family of fermions. The gluon, the photon, the W and the Z carry spin 1 and are bosons. The Higgs boson has zero spin and, needless to say, is a boson. The two faces of the Janus-like superparticle, which correspond to two particles of different spin, are always one boson and one fermion.

The symmetry of superspace is called *supersymmetry*. Supersymmetry has very special properties that make it different from any other kind of symmetry previously known. Usually symmetries either involve space transformations (like rotations or translations) or particle transformations (like the exchange between protons and neutrons). But supersymmetry is different. It relates particles with different spin, and spin is associated with rotations in physical space. Therefore supersymmetry must affect simultaneously both particle and space properties. This is the unusual feature of supersymmetry. Supersymmetry is deeply linked to the properties of space but, at the same time, it entails transformations among particles.

Real numbers, positive and negative, can be represented on a line. Algebraic operations (like addition or multiplication) transform numbers into other numbers. But the square root of a negative number was believed not to exist until 1572, when the Italian mathematician Rafael Bombelli invented imaginary numbers. Real and imaginary numbers span a plane and not simply a line. Taking the "impossible" operation of the square root of a negative number opened a new dimension in the space of numbers.

A mathematical curiosity is that if we (loosely speaking) take the square root of a translation in ordinary space – an operation that was believed not to make any sense – we obtain a translation into the new dimensions of superspace. Once again, as in the case of imaginary numbers, taking an "impossible" square root opens up new dimensions. In this case, the new dimensions are those of superspace.

The new dimensions of superspace are intimately related to the existence of spin, because spin is a necessary ingredient in the construction of supersymmetry. But particle spin is a concept foreign to classical physics, and can exist only in the world of quantum mechanics. For this reason the new dimensions of space have a quantum-mechanical nature. The coordinates of the quantum-mechanical dimensions are so unusual that they cannot even be described by ordinary numbers, but require special numbers that obey strange algebraic rules.

The idea of superspace sounds quite unusual and remarkable, and you may wonder why it might be of any relevance to our world, since every child knows that squares have areas and particles don't resemble Roman gods. But experience with spontaneous symmetry breaking has taught us to be cautious about appearances. Symmetries can sometimes deceive our simple perception, and hide inside the fundamental laws without making themselves fully manifest. As the captive scientist locked inside a room is convinced that space is not rotationally symmetric, we may blindly fail to recognize superspace around us.

In the Allegory of the Cave, Plato imagines a group of prisoners chained from birth inside a cavern and forced to eternally face a rock wall. All they can see are the shadows, cast by an enormous fire, of things that move behind their backs. In their eyes, those grey images on the wall are reality. But eventually one prisoner succeeds in breaking his chains and gets out of the cave. Only at that moment does he realize that what he has always thought to be reality is just an illusion, just a shadow of the real world.

If supersymmetry is spontaneously broken, we may live in a situation very similar to that of the prisoners inside Plato's cave. Superspace is concealed from us and we see only its shadows cast on our ordinary space. Each Janus-like superparticle actually casts not one, but two different shadows on the wall of "real" space. One shadow corresponds to a boson particle and the other one to a fermion.

Theoretical physicists imagine that the Standard Model could be a theory formulated in superspace. Quarks, leptons, and the gauge particles communicating forces constitute only half of reality. Each known particle is only one of the two shadows cast by a double-headed superparticle freely roaming in superspace. We have never

Figure 10.2 Plato's (super)cave. The two scientists cannot directly see the superparticle (here represented as a doughnut-like shape) floating in superspace, but only its two shadows (a boson and a fermion) cast on ordinary space (the wall of the cavern).

observed the second half of reality because the spontaneous breaking of supersymmetry makes the second half of superparticles too heavy for experimental detection so far. But the LHC may be the tool with which we break our chains and gain the freedom of stepping into the reality of superspace.

No doubt supersymmetry is an unusual and remarkable concept, but you may still wonder why it should have anything to do with our world. This is indeed a good and valid question. Soon after supersymmetry was discovered, physicists started to ask precisely this question, but no one had a good answer. As Father Brown, the detective priest of many of Chesterton's stories, once said: "It isn't that they can't see the solution. It is that they can't see the problem."[3] The same was true for theoretical physicists in the early days of supersymmetry. The theory was elegant and attractive, but it wasn't clear how it could be used. Supersymmetry was the solution, but nobody knew what the problem was.

Supersymmetry has had a peculiar history since the very beginning because, in the early 1970s, it was discovered not once but three times. The French physicist Pierre Ramond, later in collaboration with André Neveu and John Schwarz, made the discovery, but in a rather abstract context, and the connection with the world of particles was not evident. At about the same time, Yuri Golfand and Evgeny Likhtman, and later Dmitri Volkov and Vladimir Akulov, discovered supersymmetry in the Soviet Union, but the idea remained hidden behind the Iron Curtain. Finally, the seminal work by Julius Wess and Bruno Zumino decisively established the interest of a large number of physicists in the idea.

At the beginning, supersymmetry was studied mostly for the sake of pure theoretical speculation, and some people found such activity objectionable. This attitude is illustrated by two incidents that occurred to Michael Duff when he was a lecturer at Imperial College, London. As Gordon Kane relates: "In 1979 the theory group there applied for funding to support research activities, particularly postdocs. Their request was approved, contingent on the funds *not* being spent on supersymmetry research. A couple of years later, Duff applied for support to attend a meeting on supergravity that Stephen Hawking was organizing in Cambridge. The request was rejected, with the explanation that such research was not deemed a suitable use for funds in particle theory."[4]

Theoretical physics is powerful and effective especially when it is left to wander freely in the realm of unbridled speculations. Of course most of the ideas generated in this process end up in blind alleys, but it is sufficient that only one of them hits the right target and progress is

[3] G.K. Chesterton, *The Point of a Pin*, in *The Scandal of Father Brown*, Cassell, London 1935.

[4] G. Kane, *Supersymmetry*, Perseus Publishing, Cambridge 2000.

made. Groundbreaking ideas rarely occur when theoretical physicists follow established routes, but instead sprout from the liberty of pursuing instinctive intuitions. Supersymmetry was such an appealing concept that it was hard to concede that nature did not exploit the harmony of this new kind of symmetry.

Supersymmetry and naturalness

> I can't understand why people are frightened of new ideas. I'm frightened of the old ones.
>
> John Cage[5]

Eventually such thinking was not in vain. A good problem for which supersymmetry could be the solution was identified at the beginning of the 1980s. Supersymmetry can solve the naturalness problem because virtual particles behave in a much more disciplined way in supersymmetric theories, as they are restrained by a symmetry principle, which is the only language they listen to. To understand how supersymmetry could cure the naturalness problem, let us proceed with an analogy.

It is your son's birthday and you organize a party in which all of his schoolmates are invited. To have a more festive atmosphere, you carefully arrange in a neat order colourful balloons spread uniformly around the house. But as soon as the invited children arrive, absolute chaos breaks loose. There is no way for you to control the crowd of energetic kids who run ceaselessly all over the house, bumping into each piece of furniture, kicking everything they find in their way, and pushing things around. Your neat and uniform arrangement of balloons is instantly destroyed. Balloons fly everywhere, bouncing from one corner of the house to the opposite one in a matter of seconds.

The unruly kids are like virtual particles, and the uniform arrangement of balloons is the Higgs substance. Virtual particles communicate their energy to the Higgs substance, which is disrupted and becomes denser, making the weak length shorter. We are facing the naturalness problem.

Defeated by your futile attempts to bring order to the chaotic party, you collapse in an armchair, where you fall into a deep sleep and start dreaming. In your dream each child metamorphoses into a little Janus. And then something even more extraordinary happens. Every time one side of the child tries to kick a balloon, the other side simultaneously responds by giving an equal kick in the opposite direction. The two

---

[5] J. Cage, as quoted in R. Kostelanetz, *Conversing with Cage*, Routledge, New York 2003.

kicks exactly compensate and miraculously leave every balloon perfectly still. The little Januses keep on running amok around the house as wild as ever, but your neat arrangement of balloons remains undisturbed in its original uniform distribution. Still fast asleep, you smile, pleased by this comforting dream.

Supersymmetry solves the naturalness problem in a very similar way. Each of the two particles forming a superparticle gives a large contribution to the density of the Higgs substance. The two contributions are exactly equal, but with opposite sign, and so they cancel each other out, just as the two opposite kicks of the little Januses neutralize their total effect. When one performs the actual calculation for the first time, this perfect cancellation between large contributions looks like a miracle. But it is not a fortuitous accident; it is a manifestation of the power of symmetries. The Higgs substance, which in ordinary space is disrupted by virtual particles, is left perfectly undisturbed in the vast splendour of superspace. The frenzy of virtual superparticles does not affect the Higgs substance, because thus it is written in the laws of symmetry.

The spontaneous breaking of supersymmetry slightly modifies the double identity of the superparticle, making one of the two particles heavier than the other. As a result, the cancellation of the effects of virtual particles in superspace is only nearly exact. The requirement that the remaining effect does not reintroduce a naturalness problem leads to the conclusion that the new particles predicted by supersymmetry must have masses smaller than about 1 TeV, which is well within the territory of zeptospace. This is the reason why supersymmetry fans are so excited about the prospects of the LHC. Supersymmetry predicts that every particle of the Standard Model has a more massive duplicate, which must be within the reach of the LHC. If the theory is true, the LHC will discover that zeptospace is nothing other than a form of superspace.

Supersymmetry and unification

> The ground of physics is littered with the corpses of unified theories.
>
> Freeman Dyson[6]

The theoretical investigation of supersymmetry is motivated not only by the naturalness problem, but also by the quest for further unification. Superparticles have the dual identity of spin-½ particles (like quarks and leptons) and spin-1 particles (like gluons, photons, W, and Z). For

---

[6] F. Dyson, *Disturbing the Universe*, Harper & Row, New York 1979.

this reason, it is sometimes said that supersymmetry unifies the concepts of matter and force. However, matter particles and force carriers are not part of the same superparticle and thus there is no further unification between force and matter beyond what is already achieved in ordinary field theory. Supersymmetry elevates the Standard Model to superspace, but maintains the same gauge structure in the description of forces. Even so, supersymmetry may have a lot to do with unification of forces.

We have already mentioned how supersymmetry is intertwined with the properties of space. But, as revealed by Einstein's general relativity, the properties of space are related to the force of gravity. In fact, gravity finds itself automatically integrated into a new theory where supersymmetry acts as a local symmetry rather than as a global one. This theory, first identified in 1976 by Sergio Ferrara, Daniel Freedman, and Peter van Nieuwenhuizen, has the very appropriate name of *supergravity*. Supergravity provides a potential link between gravity and the other forces.

The hardest problem in the unification of gravity with the other gauge forces is the reconciliation between general relativity and quantum mechanics. The only known theory that achieves this task in a single consistent framework is *string theory*. In string theory there are no particles, but tiny extended objects – called strings – that propagate in space and oscillate, creating vibrations that we detect as particles. At present, vigorous efforts are being made in attempts to understand the complexity of the theory, and there are indications that string theory may represent a higher (if not the ultimate) step in Jacob's ladder.

Supersymmetry comes in because it is a necessary ingredient of a consistent string theory, which, for this reason, is often called *superstring theory*. So supersymmetry could indeed be a key element in nature's design. Nevertheless this argument by itself is not a sufficient reason to expect the discovery of supersymmetry at the LHC. Superstrings, if they exist, are probably entities belonging to a world much more remote than zeptospace, well beyond the grasp of the LHC.

There is yet another context in which supersymmetry could play an important role in the quest for unification. Long before supersymmetry became part of the toolkit of physicists, Howard Georgi and Sheldon Glashow suggested that strong and electroweak forces could be different facets of a single force. Their proposal was emphatically called *grand unified theory*.

At first sight the suggestion seems utterly preposterous. To understand why the proposal of grand unification sounds impracticable, let us recall the structure of gauge theory. It is a rather intuitive concept that symmetries identify certain geometrical forms. For instance, rotational symmetry in space identifies the sphere, because the sphere is the only (simply connected) solid completely invariant under rotations. The form of the solid is fully determined by the symmetry principle alone; only one number – the radius of the sphere – remains arbitrary. Although less intuitive, exactly the same thing happens in gauge theory. Gauge symmetry fully determines the structure of particle interactions, save for one number. This number, called the *coupling constant*, measures the strength of the gauge force. Everything else is law carved in stone by the chisel of symmetry.

The Standard Model describes strong, weak, and electromagnetic forces in terms of the gauge symmetry corresponding to the product of three Lie groups, and therefore has three coupling constants. These three coupling constants determine the strength of the strong, weak, and electromagnetic forces and have been measured very accurately. Not surprisingly, the coupling constant of the strong force is larger than that of the weak force. This raises an immediate objection to the idea of grand unification: how can a single gauge force, which contains only one coupling constant, describe three forces of different strengths?

The answer to this question lies in quantum mechanics and, more precisely, in the presence of the ubiquitous virtual particles. Let me start by considering the electromagnetic force. According to classical physics, an electrically charged object – say, an electron – exerts a force whose strength increases, as we get closer, with the square of the distance. But things become more complicated in quantum mechanics. Space is infested with insubordinate gangs of virtual particles that immediately notice the presence of the electron. The electron repels negatively charged virtual particles, while attracting positively charged ones. As a result, the electron is surrounded by a swarm of positive charges. The cloud of positively charged virtual particles, constantly appearing and disappearing, effectively reduces the strength of the electron charge when measured from far away. As we get closer to the electron, we measure a stronger charge because the cloud of virtual particles in front of us is reduced. The effect is akin to gazing at a street light through the fog on a winter Geneva evening. The fog dims the light, but the dimming is reduced as we approach the lamp-post.

This phenomenon gives rise to a singular result. In classical physics, the electromagnetic force depends on the distance, but the electric charge is a constant number. Instead, in quantum mechanics, the electric charge depends on the distance at which we observe it, because it is screened by the cloud of virtual particles. And the surprises aren't over.

Let us extend our considerations from electromagnetism (QED) to the strong force (QCD). The electric charge plays the role of the coupling constant of QED, while QCD has a different coupling constant, which describes the strength of the strong force. As you might expect, in QCD also, the coupling constant depends on the distance at which the force is probed. But, as we move away from a quark, the coupling constant of

QCD increases, rather than decreasing as in the case of QED, because virtual gluons reinforce the QCD charge rather than screening it. The fog of QCD virtual particles makes the street light brighter, as we move away from it. This unexpected result earned a Nobel Prize for Gross, Politzer, and Wilczek and paved the way to understanding the strong force in terms of QCD, as described in Chapter 4.

Coupling constants are not constant, but vary with the distance at which they are probed. This may raise some objections to the congruity of the terminology, but physics is notoriously full of constants that vary. The variation of coupling constants with distance is the key for the unification of gauge forces.

Let us take our usual virtual starship of theoretical calculations and embark on a journey towards the depth of the unimaginably small. We have just seen how, as we move towards smaller distances, the coupling constant of QCD becomes smaller while the one of QED becomes larger. So, as we proceed in our journey, we discover that the strong force becomes weaker and the electromagnetic force stronger. The three coupling constants of the Standard Model become progressively more similar. By the time we reach a distance of about 10^{-32} m, the so-called *grand-unified length*, the three coupling constants become nearly equal.

Conceiving physical reality at the grand-unified length requires a gigantic leap of imagination. The grand-unified length is equivalent to 10^{-11} zeptometres, and this is way beyond what can be explored at the LHC. Trying to identify the grand-unified length at the LHC is like using regular binoculars to spot molecules on the surface of the moon.

It is stunningly remarkable that the strengths of strong, weak, and electromagnetic forces become nearly equal at the grand-unified length. This empirical fact suggests that at those sensationally small distances forces merge into a single entity. Georgi and Glashow suggested that this entity is a grand unified theory, which is a gauge theory described by the symmetry of a single Lie group, containing a single coupling constant. In a grand unified theory not only strong, weak, and electro-magnetic forces become different facets of a single force, but also quarks and leptons can be viewed as different facets of unified particles. The unification of forces achieves a partial unification of matter too. The idea of grand unification, albeit confined in a world far from direct experimental exploration, is indeed fascinating.

Today the experimental measurements on the three coupling constants of the Standard Model are much more precise than the time at which grand unified theories were first proposed. When these measure-ments (relative to a distance of about two billionths of nanometre) are extrapolated through theoretical calculations to distances of about 10^{-32} m, the three coupling constants almost meet at a point, but not quite, as shown in the top frame of Figure 10.3. This approximate agree-ment could already be viewed as a good indication in favour of grand unification. After all, our knowledge of the world at 10^{-32} m is certainly

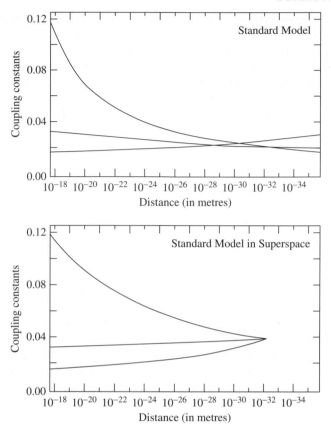

Figure 10.3 The three coupling constants of the strong, weak and electromagnetic forces as functions of the distance at which the forces are probed. Experimental measurements of the coupling constants refer to a distance of about 10^{-18} metres, while extrapolations to smaller distances are based on theoretical calculations. The top frame shows the case of the Standard Model in ordinary space, while the bottom frame shows the case of the Standard Model in superspace.

sparse and some still unknown ingredients could give the necessary correction to achieve a perfect unification of the strengths of forces.

But here comes the unexpected surprise. If the calculation that extrapolates the coupling constants is done in superspace rather than in ordinary space, the three coupling constants magically meet at a single point, within the margins of experimental uncertainties, as shown in the bottom frame of Figure 10.3. Supersymmetry is exactly the needed ingredient to merge all forces into a single unified entity. This result has stirred a lot of excitement among physicists, because it can be interpreted as a sign that supersymmetry is part of nature. Of course, the exact merging of the three coupling constants could well be a numerical coincidence, a cruel prank played by nature at the expense of gullible theoretical physicists. But, as Miss Marple once said: "Any coincidence is always worth noticing. You can always discard it later if it is only a coincidence."[7]

The accuracy with which strong, weak, and electromagnetic forces blend into a single entity is so striking in the case of supersymmetry that it is difficult to disregard this tantalizing clue. If it isn't merely a numerical coincidence, then it is the signal heralding the approaching revolution. Crossing the gate of zeptospace, the LHC will break into superspace territory where, deep inside, lies the treasure of the grand unification of forces.

Discovering supersymmetry at the LHC

> I do not seek. I find.
>
> Pablo Picasso[8]

If the ideas regarding supersymmetry are correct, for each particle of the Standard Model there exists a double, a new particle with different spin. Quarks and leptons have their boson counterparts, called *squarks* and *sleptons*. Gluons have fermion counterparts, called *gluinos*. Because of a complication related to the simultaneous breaking of supersymmetry and gauge symmetry, the fermion counterparts of the *W*, *Z*, photon, and Higgs bosons are collectively called *charginos* (if they carry electric charge) and *neutralinos* (if they are neutral).

The names given to the supersymmetric particles hardly match the elegance of the theory. But physics nomenclature undergoes waves of fashion. Until World War II, the names given to the particles displayed a sense of tradition reflected in their classical Greek derivation (proton, photon, meson, baryon, hadron, lepton,...). Then the tradition was broken, especially by Gell-Mann, and names in particle physics became more imaginative and fanciful (strangeness, quark, colour, charm, beauty, ghost,...). Later, the choice of names sometimes degenerated into a playful search for witty puns. For instance TOE and GUTs are not parts of the body but stand for the Theory Of Everything and Grand Unified Theories; we will also encounter Technicolour and WIMPs. The zoo of particle theories is full of weird beasts with improbable names. But, "what's in a name? That which we call a rose, by any other word would smell as sweet."[9] So let's leave the nomenclature and return to supersymmetry.

[7] A. Christie, *Miss Marple: The Complete Short Stories*, Penguin Putnam, New York 1985.

[8] P. Picasso, interview to *The Art*, 25 May 1923.

[9] W. Shakespeare, *Romeo and Juliet*, Act II, Scene II.

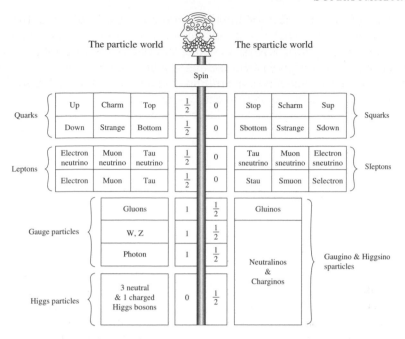

Figure 10.4 The particle content of the Standard Model in superspace.

The doubling of particles imposed by supersymmetry is akin to Dirac's prediction that each particle must have an antiparticle counterpart. Antimatter, which is associated with an exact symmetry, must have precisely the same mass as matter. In contrast, supersymmetric particles are heavier than ordinary Standard Model particles, for supersymmetry is a broken symmetry. We ignore how heavy supersymmetric particles are, because we do not know what is the exact mechanism of supersymmetry breaking. But the resolution of the naturalness problem gives us hope that supersymmetric particles are light enough to be created by proton collisions at the LHC.

The proliferation of particles predicted by supersymmetry may appear, at first sight, as a complication rather than a simplification of the theory. But this impression is just the illusion of the multiple shadows cast on the wall of the cave where we are kept prisoners. Only by staring directly at superspace, can we recognize the beauty of symmetry in the world of particles.

Discovering supersymmetry at the LHC means finding the missing half of the superworld, the second face of Janus. It is like exploring the dark side of the moon and finding that the moon is a sphere and not just a disk in the sky. But the exact way in which supersymmetry can be identified at the LHC is still a much debated subject among theorists.

One of the most favoured options in the market of ideas is that all supersymmetric particles promptly decay, save for the lightest of them, a neutralino, which is a massive, stable, and electrically neutral particle. The neutralino resembles the ordinary neutrino, although it is much heavier. Just like neutrinos, neutralinos too are invisible to LHC detectors and go through the instruments without leaving a single trace. But experiments are able to detect the presence of the neutralino from its "absence". Let me explain the concept with an analogy.

Imagine one of those automatic change machines, where you insert a 10-euro note and 10 tinkling euro coins fall into your hand. But suppose that after inserting a note into the slot you count only nine coins. The logical conclusion is: either the machine doesn't work, or you didn't catch one of the coins as they were falling into your hand.

Similarly, suppose that the sum of the energies of all particles produced in a collision at the LHC does not match the energies of the incoming protons. Energy, like money (in principle), is conserved. Thus, either the LHC detector doesn't work properly and mismeasures the energy of particles, or the detector doesn't register the passing through of some "invisible" particle. Once a thorough calibration of all instruments has been done and the performance of the detector is well understood, one is left with the second option. An invisible particle was produced in the collision and went through the apparatus perfectly undetected. Thus, a new weakly-interacting particle can be discovered through its "absence".

In more technical words, experiments observe an imbalance of particles in collision events that apparently violate conservation of energy and momentum. The "missing energy", which re-establishes energy conservation, is then attributed to some undetected particle, possibly a neutralino. In practice, the energy of particles moving straight along the proton beam cannot be measured and thus experiments must rely only on the particle motion orthogonal to the beam. But this doesn't really change the concept.

The search for "missing energy" is suggestive of Pauli's deduction of the existence of neutrinos (presented in Chapter 3), although here the situation is reversed. Pauli knew the phenomenon and found the explanation. Here we have the explanation and we are looking for the phenomenon. This is what happens when theory runs ahead of experiment.

"Missing energy" is not the only way in which supersymmetry could show up at the LHC, and theoretical physicists have proposed many other possibilities. Supersymmetric theories are like chameleons and can take many forms when confronted with experiments. For this reason, some physicists jokingly say that "supersymmetry will certainly be discovered at the LHC, even if it doesn't exist!" The point is that, in one of its mutant forms, supersymmetry could explain almost any unusual signal observed at the LHC. Sarcasm aside, I am confident that experimentalists, after cautious and attentive analysis of LHC data, will be able to sift through the various theoretical alternatives and state with certainty whether superspace is in sight or not.

A very attractive feature of supersymmetry is that it embeds the Higgs mechanism inside the gauge theory in an interesting way. The Higgs substance emerges automatically with no need to introduce special self-interactions among Higgs fields, as done in the Standard Model. For this reason, the Higgs mass can be computed in supersymmetry. The theory predicts that the Higgs boson mass must be smaller than about 120 or 130 GeV, depending on specific assumptions. This result provides a crucial experimental test for supersymmetry at the LHC. Moreover, in supersymmetry there is not only one kind of Higgs boson, but four of them: one electrically charged and three neutral particles.

Supersymmetric theories predict enough new particles to keep experimentalists busy until retirement age. But the excitement stirred by supersymmetry is not about the discovery of some unknown particles. It is about a revolutionary concept of symmetry, about the realization that space has new quantum dimensions. Things are not only "here" and "now" in superspace, but their positions are determined also by other coordinates that cannot even be described by ordinary numbers. The discovery of supersymmetry would be one of the greatest intellectual revolutions in our understanding of the structure of space.

11

From Extra Dimensions to New Forces

==>∞∞∞==

Nihil tam absurde dici potest quod non dicatur ab aliquo philoso-
phorum.

[Nothing too absurd can be said that it has not been said by some philosopher.]

Marcus Tullius Cicero[1]

Unknown worlds hidden in new dimensions of space are an all-time favourite of science fiction writers. But the notion that space could extend beyond sensory experience is as ancient as human thought. Almost all religions assert the existence of worlds inaccessible to humanity, identified either as the abode of gods, or as some form of underworld. Classical literature abounds with references to parallel worlds.

At the turn of the 20th century, the intellectual fascination with new dimensions developed more sophisticated aspects. Sigmund Freud explored the concealed dimensions of the subconscious. Cubist and surrealist painters tried to capture the essence of extra dimensions on the flat canvas. H.G. Wells promoted a new literary genre, in which the mathematical aspects of new dimensions are often present. Edwin Abbott wrote the famous novella *Flatland*, in which an imaginary two-dimensional creature visits three-dimensional space and, learning about its marvels, starts dreaming about worlds with more than three dimensions. The story is an allegory of Victorian society, but it also serves as an interesting mental exercise for dealing with the concept of different spatial dimensions.

Not surprisingly, this interest in the notion of extra dimensions coincided with a renewed scientific understanding of space and time. Around that period, relativity shook our views on these concepts. We experience a difference between space and time as our consciousness moves ceaselessly along the direction of time. But according to relativity, the physical entity in which natural phenomena occur is unified four-dimensional space-time.

[1] M.T. Cicero, *De Divinatione,* Liber II, 119.

A little simplification will help us to understand how time can be viewed as a new dimension. Borrowing Abbott's analogy, let us imagine a flat creature, a Flatlander, living in a two-dimensional space, say a sheet of paper. Now picture a sphere moving in the three-dimensional space and passing through the paper. The Flatlander is unable to see beyond his flat space and he will perceive only the part of the sphere that intersects the sheet of paper. Thus, as the sphere passes by, he will see a circle first expanding, then contracting, and finally disappearing. In other words, the Flatlander perceives a three-dimensional sphere as a two-dimensional circle changing shape in time. Similarly, we perceive four-dimensional reality as three-dimensional objects changing in time. But time is, in all respects, another dimension. Stacking the frames of a movie picture one on top of the other is a way of representing time as a vertical space direction.

Adding new dimensions to space-time is a simple mathematical exercise. Actually, from the physics point of view, adding new time dimensions leads to confusing and paradoxical results, so we will limit ourselves to new spatial dimensions. We can proceed in steps. Joining two points makes a one-dimensional segment. Joining four segments makes a two-dimensional square. Joining six squares makes a three-dimensional cube. Joining eight cubes makes a four-dimensional hyper-cube. And so on.

The mathematics is simple, but the visualization beyond three dimensions is not. The best we can do is to unfold the hypercube and project it in three dimensions, in the same way as we unfold a cube on a plane or a square on a line (see Figure 11.1). Incidentally, Salvador Dalí portrayed this unfolded hypercube in his painting *Crucifixion*. If you have a very flexible mind, you can join together the open faces of the eight cubes of the unfolded hypercube and reconstruct the four-dimensional entity. Good luck.

While mathematics is rather casual about the number of dimensions, physics can be quite different in spaces with new dimensions. For instance, the sudden disappearance of an object and its reappearance in a different place would unmistakably look like a supernatural event to us. Also, you can never twist your right hand and make it look identical to your left hand, no matter how you contort your arm. And yet, these oddities can occur if space has a new dimension. Let us see how this happens in Flatland.

Just pick up a pen that is lying on your desk and put it back on the desk few inches away. A poor Flatlander, living on the surface of the desk, will see the pen suddenly disappear from his view and reappear elsewhere. Unaware of the existence of a vertical direction, he won't believe his eyes. Also, for a Flatlander there is no way in which a right-bending arrow (↦) can be turned into a left-bending arrow (↤). Just try to rotate the arrows on a piece of paper, and you will understand the

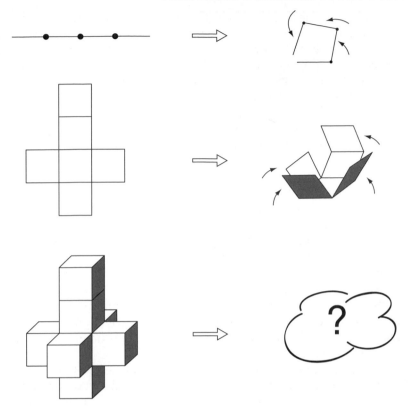

Figure 11.1 Constructing shapes in spaces with progressively higher dimensions.

difficulty experienced by a Flatlander. But if you pick up one arrow, turn it over, and then put it back on the desk, you can easily superimpose the two arrows. Similarly, a four-dimensional creature could take your right hand into his space and easily turn it into a left hand without you feeling any pain.

Extra dimensions can hide the unexpected. And there is nothing more pleasing to the mind of a theoretical physicist than the unexpected. So it is natural to consider whether extra dimensions can be part of reality. Of course we have to confront the fact that we simply do not experience the presence of any unknown space-time directions other than forwards, upwards, sideways, and later. Thus, the first issue to be addressed is: if extra spatial dimensions exist, where are they hidden? C.S. Lewis, in his saga *The Chronicles of Narnia*, hid the new space behind the back of an old wardrobe. Science must be more explicit.

Suppose that a very tiny insect from Flatland goes on holiday at a renowned Swiss ski resort. Once there, he starts crawling on the cable

Figure 11.2 From the point of view of the bug, the cable is a two-dimensional surface. For a distant observer, the cable is a one-dimensional line.

stretched between the cable-car station in town and that on the top of a mountain. From his point of view, the surface of the cable is a two-dimensional world. But there is a difference between the two directions of space. While the direction along the cable is flat, the direction around the cable is curled-up, because the Flatlandish bug can get back to the same point by walking always forward in that direction. But the insect is so small that it hardly feels any difference between the two directions – just as we usually do not even realize that the surface of the earth is curved. Consider now the point of view of a skier on the ski slopes, far from the cable. To the skier, the cable appears as a line in the sky. From that distance, the skier perceives the cable as a one-dimensional space, because his eyes are not able to resolve its thickness.

This example shows that the same space can appear to different observers as having different dimensions. Just as the second dimension of the cable is visible to the insect but invisible to the skier, physical space could have more than three dimensions, but some of them may remain hidden to our sensory experience and to our scientific instruments, because they curl up at minute distances. Only to an observer as small as the tiny bug would space reveal its true nature. In physics, the process in which spatial dimensions bundle up in small regions is called *compactification*. A world with *compactified dimensions* contains some hidden space, visible only at very short distances. So, in the previous example, we can say that one of the two dimensions of the surface of the cable, the one wrapped around it, is compactified in a circle, because a section of the cable has the geometric shape of a circle.

The assumption that certain dimensions of physical space are compactified may appear rather contrived, just an artificial way of hiding the unexpected. But general relativity has taught us that space-time is not a static and immutable entity. On the contrary, space-time bends, stretches, and contracts. Astronomical observations have proved that galaxies recede from us with a velocity roughly proportional to their distance. This global flight of galaxies away from us does not arise from a peculiar motion of galaxies with respect to the Milky Way, but from the stretching of the fabric of space. Even at this very moment, our universe is expanding. It is as if we were living inside the dough of a cake that is rising.

Moreover, as I will describe in Chapter 12, scientists believe that, at its very beginning, the universe underwent a period of dramatically rapid expansion, called *inflation*. All the space that we see with our most powerful telescopes evolved in an exceedingly short time from a space smaller than a speck of dust. Isn't it then possible that, while our three dimensions of space have expanded enormously, other spatial dimensions have remained small or even counteracted by contracting into miniscule regions? In this context, the hypothesis of compactified dimensions is rendered much more plausible.

The idea that extra spatial dimensions may shroud the secret of force unification has circulated in physics since the time relativity was born. In 1921 the mathematician Theodor Kaluza (1885–1954) proposed an inspiring idea in a work later revamped by the physicist Oskar Klein (1894–1977). The theory studied by the two scientists, now known as *Kaluza–Klein theory*, is simply general relativity formulated in five-dimensional space-time. Gravity is the only force described by the theory. But suppose that the extra spatial dimension is compactified in a small circle. Just as the skier sees the cable as a one-dimensional line, an observer unable to resolve distances as small as the radius of the compactified dimension would perceive reality as four-dimensional space-time. The surprise is that reality will appear to this observer as a world where two forces are present: gravity and electromagnetism.

The result is really astounding. According to Kaluza–Klein theory, electromagnetism is like a mirage projected by a hidden extra dimension. Electromagnetism is just the illusion that we perceive when we stare at the force of gravity without enough visual resolution to distinguish that space extends into yet another direction. The secret for this miracle lies in the symmetry. The symmetry of five-dimensional space-time corresponding to translations along the extra dimension remains hidden from us. But this hidden symmetry is exactly the gauge symmetry of QED. Thus, the gravitational force of the hidden space appears to us as the electromagnetic force.

Unfortunately, for a series of technical reasons, Kaluza–Klein theory is inconsistent with a proper description of the real world. In spite of this failure, some of the features of the theory are so compelling that they have inspired many subsequent attempts to unify forces in spaces with extra dimensions. The idea of Kaluza–Klein theory is still interesting today and is periodically resuscitated in other contexts, such as super-gravity and string theory. However, most physicists believed that the effects of extra dimensions could take place only at distances much smaller than those probed by collider experiments. The situation changed in 1998.

Large extra dimensions

> There is a fifth dimension beyond those known to man. This is the dimension of imagination.
>
> Rod Serling[2]

Nima Arkani-Hamed, Savas Dimopoulos, and Georgi Dvali were pondering together over the naturalness problem in 1998, when they decided to address it from a different point of view. Instead of looking for symmetries that could justify the largeness of the hierarchy between the weak and Planck lengths, they assumed that, in the depth of zepto-space, these two lengths are actually the same. The naturalness problem simply vanishes because there is no hierarchy to start with. But the hier-archy reflects the empirical fact that gravity is much more feeble than the weak force. Therefore, the hypothesis that no hierarchy exists in zeptospace is just a way of recasting the original problem into a new one. The new question is to understand why gravity appears so feeble to us, but not to elementary particles in zeptospace.

Since Newton's time it has been known that the force of gravity decreases as the square of distance. But this result is valid only for a space with three dimensions. If space had extra dimensions, the force of gravity would decrease much faster. For instance, had Newton lived in a world with four spatial dimensions, he would have discovered that gravity decreased with the cube of the distance. In five dimensions, he would have found the law of the fourth power of the distance, and so on. It is not difficult to understand this result. Consider the gravitational field gener-ated by a compact mass, say the sun. The more numerous the dimensions of space are, the more the gravitational field gets diluted as we move away from the sun. Hence, the strength of gravity diminishes with distance more rapidly in hyperspace – a space with extra dimensions. As

---

[2] R. Serling, preamble to the TV series *The Twilight Zone*.

an analogy, compare the spraying of water from a sprinkler with the jet of water from a hose. The density of water decreases faster with distance when water is diffused in more directions.

Turning the previous argument around we discover that, as we approach smaller distances, gravity grows stronger more rapidly if space has extra dimensions. Suppose that reality hides some compactified dimensions and imagine embarking on the virtual starship that could travel towards smaller distances. As we zoom towards microscopic depths, suddenly space opens up in new directions and, consequently, gravity becomes stronger more rapidly. By the time we reach zeptospace, the strength of gravity could catch up with the other gauge forces. The Planck length, which is a measure of the intensity of the gravitational force, could become roughly equal to the weak length, in the multi-dimensional land of zeptospace.

This proposal is a daring idea. It entails the hypothesis that we live in a space with three dimensions but, just like Flatlanders, we are unaware of the multi-dimensional space that surrounds us. This extra space is an empty and desolate place, not accessible to any of the Standard Model particles. Gravity, however, is different and is allowed in this inhospitable space. Diluting most of its strength into the vastness of this empty space, gravity appears to our senses as the weakest of all forces. Thus the hierarchy between weak and gravitational forces is merely an illusion. Gravity appears weak to us only because its strength is depleted by the vast extra-dimensional space.

The large diluting factor necessary to dispose of the hierarchy requires the existence of a large volume of space in extra dimensions. For this reason this theoretical proposal is known as *large extra dimensions*. Of course, the word "large" is relative to the typical distances of particle physics. But experimental information on the properties of gravity at small distances is so scarce that spaces that only gravity can sneak into could be surprisingly large. In 1998, the size of the compactified dimensions could have been as large as a millimetre. In the meantime, new experiments have tested gravity at small distances with improved accuracy, and today we know that the size of this hypothetical space must necessarily be smaller than about 50 microns. This is still a very large space by the standards of elementary particles.

Warped extra dimensions

Man's mind, once stretched by a new idea, never regains its original dimensions.

Oliver Wendell Holmes[3]

---

[3] O.W. Holmes, Sr., *The Autocrat of the Breakfast Table,* The Atlantic Monthly, Boston 1858.

Living like Flatlanders, caged in a three-dimensional world, while there is so much inaccessible space around us may sound like a dismal fate. But a partial consolation comes from the realization that this situation is rather commonplace in the context of string theory. In 1995 Joseph Polchinski discovered that string theory is populated with *branes*. Branes are entities with fewer dimensions than the space in which they are embedded, which are capable of confining matter and forces to their interior. The name is reminiscent of "membrane" because we can picture them as membranes floating in space, or as infinitely large sheets suspended in mid air. But since string theory is usually formulated in 10-dimensional space-time, branes can have more than the two dimensions of ordinary membranes. Thus they can describe three-dimensional worlds, like ours, suspended in a vast multi-dimensional hyperspace. The word "brane" also provides rich opportunities for making trite jokes such as remarking that anyone not working in string theory is a "physicist with no brane".

A remarkable property of branes is that they automatically provide the cage in which particles can be imprisoned. Thus, worlds suspended in a vast multi-dimensional universe, which is penetrable only to the gravitational force, are ordinary facts of life in string theory. The existence of branes and the solution of the naturalness problem suggested by large extra dimensions fuelled an extraordinary interest in the subject on the part of theoretical physicists.

A new important result came in 1999, when Lisa Randall and Raman Sundrum proposed an alternative solution to the naturalness problem, inspired by the brane world. The idea is analogous to a well-known physical phenomenon called *gravitational red shift*. Let me explain how it works. (If you have ever heard of the Doppler red shift, don't get confused. The gravitational red shift has nothing to do with the Doppler effect.)

In the Newtonian theory, mass is the only agent capable of exerting gravitational force. But in general relativity any form of energy, and not only mass, feels the effect of gravity. This result is not surprising, once we know that mass itself is nothing else but a form of energy (recall $E = mc^2$). But this assertion represented a clear departure from the classical theory. For this reason the observation during the solar eclipse of 1919 that light rays, which carry energy but no mass, are bent by the solar gravitational field provided the decisive confirmation of general relativity.

Imagine a source of a strong gravitational field, such as a very massive star. When a light ray is emitted by the star, it must expend some energy to overcome the gravitational attraction and escape from the star. This is exactly what happens when you throw a stone straight up in the air. The stone loses kinetic energy as it flies higher and, consequently, slows down. But light cannot slow down because, as special relativity teaches us, its speed remains always constant. Wavelength, and not velocity, is a measure of the energy of light; less energetic light rays have longer wavelengths. So when the light ray escapes from the massive star, it loses energy and its wavelength increases. Thus, the wavelength of a light ray is smaller when emitted from a massive star than when measured by a telescope on earth. The wavelength increases during its escape from the star. This effect is called gravitational red shift.

Randall and Sundrum imagined a similar situation on a larger scale. We, the earth, and the whole universe belong to a brane suspended in hyperspace. Far from us (in a direction which is neither forward, nor upward, nor sideways) there is another brane, which acts as a strong gravitational source, just like the star of the previous example. This source of gravity is so strong that it deforms the extra dimension that separates the two branes. For this reason, this theory is referred to as *warped extra dimensions.*

Just as the wavelength of light increases as it travels from the star to the earth through the stellar gravitational field, so the Planck length increases as it is transmitted from the distant brane to our world through the exceptionally strong gravitational field of the warped dimensions. The hierarchy between the weak and Planck length is just an illusion. Gravity appears weak to us only because we see it through the distorting lens of warped extra dimensions.

This new proposal threw more gasoline onto the fire of extra dimensions raging in the heads of theoretical physicists. Many new variations of the theory were proposed. The solution of the naturalness problem based on warped extra dimensions requires the Higgs field to be confined to the brane, but it does not specify the location of the other quantum fields, like those of quarks, leptons, and the gauge particles communicating forces. Some especially interesting results were obtained when particles of the Standard Model, and not only gravity, were allowed to roam in the territory of warped extra dimensions, where the strong gravitational force distorts and disfigures many of their ordinary properties. But the most surprising result came from an unexpected relationship.

In 1997 Juan Maldacena put forward an audacious conjecture. He suggested that some theories of gravity in five-dimensional space-time are completely equivalent to certain gauge theories in ordinary four-dimensional space-time. This may superficially sound reminiscent of Kaluza–Klein theory but it is really quite different. According to Kaluza and Klein, a theory of gravity in five-dimensional space-time looks *approximately* the same as gravity *and* electromagnetism in four-dimensional space-time, as long as the observer is unable to discern the size of the hidden space. In contrast, according to this new conjecture, two theories defined in spaces with different dimensions are two *identical* descriptions of the same reality. Although the conjecture has never been proved with mathematical rigour, there is now mounting evidence that it must be correct.

This result is very surprising because gravity in extra dimensions and gauge theory in ordinary space do not seem to have much in common. The secret of the correspondence lies in a nearly magical phenomenon: holography. A hologram is the fascinating result of a clever photographic technique. As you move your head left and right in front of a hologram you can see the different sides of the object reproduced by the image, as if they were real. A hologram captures on a two-dimensional plane the full information of a three-dimensional image. Similarly, the four-dimensional gauge theory can capture all the information of gravity in five dimensions. Although superficially very different, the two theories are two facets of the same reality.

Maldacena's conjecture is suggestive of the fact that the setting of warped extra dimensions is equivalent to certain gauge theories, thus providing an unexpected connection. Long before theoretical physicists lost their "branes" in extra dimensions, Steven Weinberg and Leonard Susskind had suggested the earliest solution to the naturalness problem. The idea was to replace the Higgs field with a new gauge force, dubbed *technicolour* because of the strict analogy with the "colour" introduced in QCD. Technicolour provides a very elegant way of explaining how space can be filled with the substance that generates the masses of the W and Z particles, and it brings back the spontaneous breaking of electroweak symmetry into the edifice of the gauge principle. Later, the theory ran into difficulties when confronted with experimental data, and required some modifications. The gauge theory associated with warped extra dimensions automatically contains the elements that can possibly resolve some of the difficulties of technicolour.

This result demystifies the concept of warped extra dimensions. Quantum mechanics teaches us that it is pointless to argue whether an electron is a particle or a wave. Particles and waves are two descriptions of the same physical entity. Similarly, in some cases, there is no conceptual distinction between extra spatial dimensions and new forces. Dimensions and forces can be two different descriptions of the same physical entity.

The search for extra dimensions at the LHC

> Sometimes my dreams take me to other dimensions.
> Uri Geller[4]

---

[4] U. Geller, interview for *Holistic London Guide*, 1997.

If the picture of branes and extra dimensions sounds too much like science fiction for you to take seriously, don't worry; some physicists share this opinion. But the idea that there is another world just beyond the frontier of our knowledge is so enticing that it cannot be easily dispelled from the minds of many physicists. The most remarkable aspect of the story is that the LHC can put these fanciful theoretical imaginings under solid experimental scrutiny. If any of these hypothetical theories are true, then the LHC will discover that zeptospace is a multi-dimensional hyperspace that extends into new physical directions.

In Abbott's novel the Flatlander, upon returning to Flatland after his visit to three-dimensional space, could not find any expression to convey the notion of Spaceland other than saying: "Upward, not Northward!" These words were not sufficient to convince any of his fellow countrymen about the wonders of Spaceland and he was eventually imprisoned for his treacherous ideas. Can experimental physicists find a better way to convince the world that the LHC has visited hyperspace?

The LHC detectors cannot directly measure a physical distance in extra dimensions, but experimentalists can infer the existence of hyperspace by studying the echoes sent by particles moving in the hidden space. Let me explain the idea starting with the previous analogy of the tiny insect on a Swiss holiday. The Flatlandish bug is crawling on the surface of the cable, happily spiralling around it in its ascent towards the alpine peak. But the distant skier sees any motion on the cable as occurring on a one-dimensional line. Any spinning around the cable goes undetected.

Similarly, consider a particle moving in a space with some compactified dimensions. The particle moves along some of the flat directions of space and, at the same time, spirals around the curled-up dimensions, just as the bug does in his winding path. The LHC instruments, like the eyes of the skier, are unable to resolve the compactified dimensions, and observe the particle as if it were moving in ordinary space. But there is some kinetic energy associated with the spinning of the particle in the extra dimensions, which is then detected not as motion but as a form of intrinsic energy of the particle. We are by now sufficiently familiar with the equation $E = mc^2$ to recognize immediately this energy, like any other form of intrinsic energy, as mass. Thus, the motion of a particle inside a curled-up dimension is detected in our world simply as mass. The faster a particle spirals inside the extra dimensions, the heavier it will appear to us.

But quantum mechanics does not allow spiralling inside the curled-up dimensions with any possible energy. Just as electrons can occupy only special orbits inside the atom, so particles can spiral inside curled-up dimensions only with special values of energy. Particles moving inside curled-up dimensions are like violin strings that vibrate only in special harmonics. These harmonics are called *Kaluza–Klein modes*, and are visible in our world as particles that look perfectly identical, save for their mass. Just as a violin string emits the same note in different octaves, so each Kaluza–Klein mode has different mass but identical intrinsic properties such as charge and spin. The Kaluza–Klein modes are the echoes sent to us by a particle propagating in hyperspace. These echoes contain information about the structure of the hidden space of extra dimensions. Therefore, the detection at the LHC of various Kaluza–Klein modes would allow a reconstruction of the size and shape of hyperspace.

In conclusion, a gluon or a top quark (or any other particle) living in hyperspace would look to us perfectly identical to an ordinary gluon or top quark, but for its mass. Thus, hunting for extra dimensions at the LHC means searching for particles that look like ordinary particles, albeit they carry abnormally large masses.

The way extra dimensions are observed at the LHC helps to clarify the interpretation of Maldacena's conjecture. The strong force, described by QCD, showed up in the experiments of the 1950s and 1960s in the form of a series of new particles, called hadrons. If a new strong force like technicolour really exists, history will repeat itself in the domain of higher energies. The LHC will discover that zeptospace is full of new particles which, in analogy to QCD, are called *technihadrons*. But extra dimensions too appear in collider experiments in the form of a series of new particles – the Kaluza–Klein modes. In some cases, the properties of technihadrons turn out to be identical to those of Kaluza–Klein modes. Just as experiments cannot unambiguously determine if the electron is a particle or a wave, so the LHC will be unable to distinguish a new force like technicolour from a new warped extra dimension because, in certain cases, the experimental consequences are identical.

There is another fascinating aspect of the search for extra dimensions at the LHC. The solution of the naturalness problem based on the idea of branes implies that, at very short distances, the weak and the Planck lengths are roughly equal. If true, this means that the LHC will have the surprise of finding that in zeptospace the strength of gravity is comparable to the strength of the other forces. According to theories of extra dimensions, gravitational phenomena in zeptospace are about 10^{30} times stronger than what is expected from general relativity in ordinary space-time.

Phenomena characterized by strong gravitational effects are known to occur only in astrophysical environments where very massive and dense objects exist. But the analogue of these phenomena could occur at the LHC, if the strength of gravity is much enhanced in zeptospace. Emission of gravitational waves, gravitational deflection of quarks and gluons, and production of microscopic black holes could be experimentally measurable processes of the particle world, if these hypothetical theories are correct. Moreover, since these processes involve elementary particles and not astronomical bodies, they would enable the testing of gravity in the quantum-mechanical domain. The coexistence of gravity and quantum mechanics, which has puzzled theorists for so long, could become a subject for experimental physics.

The idea that black holes could be artificially produced in a laboratory has raised some concern in people outside the field of particle physics. However, the only kind of black holes that can possibly be produced at the LHC will evaporate in less than 10^{-26} seconds. These microscopic black holes, the size of a few hundreds of zeptometres at most, do not acquire extra mass or cause any catastrophic event for our environment during their short lifetime. Nevertheless, some people started to worry that black holes would not evaporate, although there is no theoretical framework in which this could happen.

CERN addressed seriously this public concern and in 2003 appointed a scientific committee to review the situation. The committee concluded that the hypothetical production of extra-dimensional black holes at the LHC presents no danger. Later, new experimental results and new theoretical considerations enabled particle physicists to extend and strengthen the original conclusions. A new scientific committee, in which I took part, reconsidered the situation and issued its report in 2008.

Astronomical observations completely exclude the production of any dangerous black hole at the LHC. High-energy collisions of cosmic ray particles are happening all the time, both in space and indeed on the earth. The LHC is simply reproducing, in a controlled way suitable for experimental measurements, phenomena that have been taking place for billions of years and that are still happening today all around us. Cosmic rays hitting the earth have already produced an equivalent number of collisions as a hundred thousand LHC experimental programmes, at energies equal to or higher than the LHC. On the scale of the universe, 3000 billion complete LHC experimental programmes are occurring every single second. Without any doubt, the only impact the LHC will have on the universe is to allow a huge intellectual leap for humanity.

12

Exploring the Universe
With a Microscope

⋙◦◦◦⋘

Can we actually "know" the universe? My God, it's hard enough
finding your way around in Chinatown.

Woody Allen[1]

A common misconception is that the LHC recreates the conditions of
the universe soon after the Big Bang. It does not. This belief, which
arises from confusion between the concepts of energy and temperature,
goes as follows: "The global motion of galaxies escaping from us with
a velocity proportional to distance proves that our universe is presently
expanding. Tracing cosmological history back in time, we deduce that
the universe was once very hot and dense, and thus consisted of a
primordial soup of particles. Under these conditions, particles were
carrying a lot of energy and were frequently bumping into each other,
which is exactly what is happening at the LHC. Hence, the high-energy
proton collisions at the LHC recreate the conditions of the early
universe." However, the story is not as simple as that.

Near the beginning of time, the universe was indeed made of a hot
soup of particles, forming a large thermal system. But a thermal system
develops collective phenomena that are determined by the statistical
properties of large numbers of particles and that cannot be reproduced
by individual constituents. For instance, we have no way of recreating
the transition from ice to water and to vapour by colliding two separate
H_2O molecules. The phase transitions between ice, water, and vapour
are the result of collective phenomena, which simultaneously involve a
large number of constituents. The history of the universe was largely
influenced by collective phenomena, which cannot be reproduced by
individual proton collisions at the LHC.

There is one exception. The collisions of heavy nuclei at the LHC,
studied primarily by the ALICE experiment, produce a system with a
large number of particles. For a time of about 10^{-23} seconds, this system

[1] W. Allen, *Getting Even,* First Vintage Books, New York 1978.

can be in a high-temperature and high-density state, thus reproducing conditions similar to those in the early universe. However, the energies involved in the collisions between these heavy nuclei are not large enough to probe the physical laws in zeptospace.

The ineffectiveness of proton collisions to directly recreate the primordial conditions of the universe does not mean that the LHC will have nothing to say about early cosmological history. Indeed, proton collisions at the LHC may become an essential tool for expanding our knowledge of the early stages of the universe. This is because we believe that the same laws that govern the particle world are also ultimately responsible for the evolution of the universe. Only by extending our understanding of these laws to the smallest possible distances, will we be able to address some of the most fundamental questions about the origin of the cosmos.

The connection between the world of particle physics and the structure of our universe is probably one of the most profound results of modern science, which captivates the imagination of anyone who is confronted with its wonders. Over the past decades, cosmology has progressed enormously in strengthening this connection and in revealing an ever-sharper image of the origin of the universe.

It is almost paradoxical how cosmologists, to a first approximation, neglect all forces other than gravity, while particle physicists spend their time scratching their heads perplexed by the feebleness of gravity. For very large systems, gravity is the dominant force and thus is the main driver of the evolution of the universe. But this brings up an immediate puzzle. The electrostatic force can be either attractive or repulsive, because both positive and negative charges exist. On the other hand, according to Newton, mass is the only source of gravity and, since no "negative" mass exists, gravity is always attractive. So the puzzle is: what started the expansion of the universe? Does repulsive "antigravity" exist?

The solution to this puzzle lies deep in general relativity. In Einstein's theory, any form of energy acts as a source of gravity, for mass is equivalent to energy. But the energy density of a system is always positive, and thus it exerts only attractive gravitational forces. Luckily, there is more. In contrast with the Newtonian theory, in general relativity pressure is a source of gravity too. We are familiar with the idea that a difference in pressure creates a force. For instance, the pressure of an expanding gas pushes the pistons in a car engine. The novelty of general relativity is the assertion that the existence of pressure generates a gravitational force. But pressure can act outwards or inwards or, in other words, it can be positive or negative. Thus one draws the startling conclusion that gravity can be *repulsive*. Negative pressure exerts antigravity.

Repulsive gravity can be the engine that gave the starting kick and set into motion the expansion of the universe. The question is to identify an agent with sufficiently large negative pressure capable of overcoming the gravitational attraction of all the mass and radiation in the universe. What is this powerful agent? Any form of known matter is certainly excluded, because its energy always exceeds pressure, leading to strictly attractive gravitational forces. This agent must be a very unusual substance.

In 1917 Einstein found the answer to our question, although his goal was different. Einstein was disturbed by the fact that the equations of general relativity predicted that the universe was evolving, rather than remaining static and he found this result unacceptable. At that time it was taken for granted that the universe must be static, for there was no astronomical evidence to the contrary. Little did he know that, 12 years later, Edwin Hubble would discover that the universe is indeed expanding. As Dirac later did with antimatter, Einstein was involuntarily doing his best to obtain the wrong prediction from his equation. Later in his life, Einstein confessed to George Gamow that this had been the "biggest blunder of his life"[2].

In his calculations, Einstein discovered an unconventional form of energy that can uniformly fill all space and have negative pressure, thus exerting a repulsive gravitational force that can compete with the attraction caused by matter and radiation. This substance, called the *cosmological constant*, is a form of energy whose properties are very different from those of ordinary matter. If the cosmological constant reminds you of the Higgs substance, you are on the right track.

In 1980 Alan Guth pursued this very same track and obtained an extraordinary result. Suppose that space is filled with a new quantum field which, for reasons that will soon become clear, is called the *inflaton*. The inflaton field is very much akin to the Higgs field, although its origin is even more mysterious. As in the case of the Higgs field, the inflaton substance can permeate all space. At primordial times this substance is so dense that it overcomes the effect of any other element in the universe. The inflaton substance then behaves just like the cosmological constant, and the antigravity from its enormous negative pressure drives an explosive stretching of space, expanding the universe at a prodigious rate. In a time of presumably 10^{-35} seconds, the size of the universe expanded by a factor of at least 10^{30}, and possibly by much more. This is equivalent to blowing up a 20-nanometre virus into a gigantic creature, the size of the distance from here to the galaxy of Andromeda, in the time it takes light to cross a few millionths of a zeptometre! This astounding phenomenon is called *inflation*, but even the price increase in the Weimar Republic would pale in comparison.

---

[2] G. Gamow, *My World Line: an Informal Autobiography*, Viking, New York 1970. I thank Robert Kirshner for pointing out this reference to me.

After about 10^{-35} seconds, the inflaton substance disappears from space transferring all its energy into a hot soup of ordinary particles. But in that very short time, the inflationary burst sets all the initial conditions that determine the future fate of the universe. Inflation may sound like a fanciful hypothesis bordering on science fiction, but it is indeed a solid scientific theory, which makes very detailed predictions that can be tested against direct cosmological observations of the early epoch of the universe.

The idea of "observing" the early universe may at first appear absurd, but light travels at a finite speed and therefore we see distant objects as they were in the past. We see the sun as it was about eight minutes ago; we see Andromeda as it was more than two million years ago. Because of the finite speed of light, the concept of time blends into the concept of space.

We can travel back in time by observing ever more distant objects. One may wonder whether, in this way, we could witness the very birth of the universe with a sufficiently powerful telescope. Alas, this is not possible. Just as a wall impedes our view beyond it, so astronomical observations reach the time when the universe was about 380 000 years old, but cannot go further. Before that time, the temperature in the universe was more than 2700°C and the frantic motion of particles knocked every electron off its atomic orbit. Just as light cannot go through a brick wall because it is absorbed by the material, so any form of electromagnetic radiation could not freely travel across the medium of unleashed electrons and nuclei. The universe was completely opaque. Any image of the universe prior to 380 000 years is permanently erased and cannot cross the impenetrable wall made of hot electrons and nuclei.

But that wall is painted with the earliest and most vivid image of the universe that we can possibly obtain. Staring at that image has been a favourite activity of many teams of observational cosmologists, and this activity has been rewarded with a wealth of scientific information. That image (shown in Figure 12.1) is called the *cosmic microwave background* and it is a picture of the radiation at the moment in which nuclei trapped electrons in atomic orbits and the universe suddenly became transparent to light. Since that moment, this radiation has travelled through space essentially undisturbed.

The cosmic microwave background was first discovered in 1965 by Arno Penzias (Nobel Prize 1978) and Robert Wilson (Nobel Prize 1978), although it was initially misinterpreted as the effect of the pigeon droppings (described by Penzias as "white dielectric material"[3]) covering

---

[3] A.A. Penzias, as quoted in J. Bernstein, *Three Degrees above Zero: Bell Laboratories in the Information Age*, Cambridge University Press, Cambridge 1984.

their microwave horn antenna. More recently the COBE satellite (COsmic Background Explorer), launched by NASA in 1989, has obtained very precise measurements of this radiation. COBE's results won Nobel Prizes in 2006 for its two principal investigators, George Smoot and John Mather. WMAP (Wilkinson Microwave Anisotropy Probe), a NASA space mission launched in 2001, has dramatically improved the precision in the measurements of the cosmic microwave background and has extracted from the data precious information about the early universe. In May 2009, the European Space Agency launched a new space observatory, named Planck, expected to further improve the accuracy of the measurements.

The image of the cosmic microwave background, together with other astronomical observations, has revealed some of the most convincing evidence in favour of the theory of inflation. Let me explain how this is so.

In general relativity space is a dynamical entity that curves and bends in various shapes. Thus it is perfectly legitimate to wonder about the shape of the universe. Is the space in which we live curved or flat? Inflation gives a straight answer to this question. The fierce stretching of space, which occurred during inflation, irons out any kind of irregularity or bumpiness originally present. Just as an elastic fabric becomes perfectly flat when stretched, so space in the universe, no matter how it started, will end up almost exactly flat after the period of inflation. This theoretical prediction can be confronted with data from the cosmic microwave background. Light reaching us from 380 000 years after the

Figure 12.1 The temperature fluctuations of the cosmic microwave background in the sky, as measured by WMAP. Lighter regions are slightly warmer and darker regions are slightly cooler than the average temperature of 2.725 degrees above absolute zero.
Source: NASA / WMAP Science Team.

Big Bang probes the structure of space and can be used to measure its intrinsic geometry. Any curvature of space bends light and distorts the image of the cosmic microwave background, affecting the size of the spots shown in Figure 12.1. The interpretation of data leads to the conclusion that space in the universe is *flat* with great accuracy, fully confirming the prediction of inflation.

The temperature of the cosmic microwave background today is almost perfectly equal at every point in the sky, but this remarkable uniformity brings up a puzzle. Light always propagates at a finite speed and thus, at the time when the universe was only 380 000 years old, no light ray could have possibly crossed the full sky that we observe today. This raises the question of why the temperature of the cosmic microwave background is so uniform over such a vast region of space, if no physical information could have been transmitted between distant parts of the sky when the microwave background was emitted. This problem had puzzled cosmologists for a long time, until inflation offered a natural explanation. The full sky we observe today originates from a tiny speck of space, in which all parts influenced each other, thus ensuring the conditions for a uniform temperature of the cosmic microwave background.

Inflation predicts that the universe should be uniform and homogeneous because of the exorbitant stretching of space that occurred at its very beginning. At first sight this doesn't sound like a very good prediction. We are not a population of uniform and homogeneous beings (I believe), and just staring at the night sky reveals that our cosmos is full of stars, galaxies, clusters of galaxies, and it is not a dull and stale mass of gas. On the contrary, the inflationary prediction is extremely successful. When viewed at very large distance scales, our universe indeed appears extremely uniform and homogeneous, and many astronomical observations have confirmed this result. Our position with respect to the universe is akin to the point of view of a microscopic creature, no larger than an atom, with respect to matter. That tiny creature sees matter as a lumpy substance, full of small structures, while we see matter as a uniform medium. Similarly, we see the universe as a combination of individual stars and galaxies although, at much larger distances, it appears remarkably uniform.

So inflation successfully explains the large structure of the universe. But how can it accommodate the presence of all the individual galaxies that we observe in the night sky? The answer to this question is the most amazing part of the story. Believe it or not, it all boils down to Heisenberg's principle.

Heisenberg's principle describes the intrinsic uncertainty of various physical quantities in the world of quantum mechanics. The inflaton field is no exception and it is subjected to microscopic quantum fluctuations. As a consequence, at the time of inflation, the energy stored in the inflaton field slightly varied in different places, causing some micro-scopic lumpiness. At distances so small that quantum mechanics becomes relevant, some regions of space were slightly denser than others. But when space underwent its extraordinary stretching, the microscopic lumpiness was blown up to enormous structures of astro-nomical sizes. After the inflationary period was over, the slightly denser regions started to contract, because of their attractive gravitational force, and formed galaxies. Out of the original seeds created by the quantum fluctuations of the inflaton field grew the galaxies that now shine and are visible on clear nights.

Theoretical calculations of the structures produced by inflation are in very good agreement with the astronomical observations of the distri-bution of galaxies. Moreover, a splendid confirmation of these ideas comes from the measurements of the cosmic microwave background. The temperature of this radiation today is 2.7 degrees above absolute zero, but it is not perfectly identical in various locations in the sky, varying by an average amount of about 30 millionths of a degree, as shown in Figure 12.1.

The tiny variations in the temperature of the cosmic microwave background that have been precisely measured by WMAP are explained in terms of the same process that created the seeds for the formation of galaxies. The slightly cooler or warmer regions in the cosmic micro-wave background correspond to the primordial lumpiness of the inflaton field, which is then stretched across the sky by inflation into the image that we observe today. Measurements of the cosmic microwave back-ground are in superb agreement with the inflationary hypothesis.

The picture emerging from inflation is absolutely mind-boggling. Both the pattern of the galaxies that we observe in the night sky and the thermal variations in the cosmic microwave background are the fossils of the fluctuations of a quantum field frozen in space from the moment our universe came into existence. Physics at some of the smallest imag-inable scales – those of the inflaton quantum fluctuations – determine some of the largest structures in the universe.

The study of matter requires the most powerful microscopes – in the form of particle accelerators such as the LHC – to descend into the smallest distances, where we discover the world of quantum fields. The study of the universe requires the most powerful telescopes and satellite instruments to view in the sky the image of a quantum field blown up out of all proportion. The same elements and the same universal laws of physics describe the very small and the very large.

It is most unlikely that the LHC will give us a direct answer about the origin of the inflaton field or about many of the other still open fundamental questions in cosmology. However, it is by now clear that those answers have to be looked for within a fundamental theory of particle physics superseding the Standard Model, and the LHC can provide the clues necessary to uncover such a theory. In this sense, the LHC may contribute to solving some of the mysteries regarding the origin of the cosmos.

However, there is one issue related to the structure of the universe for which the LHC could provide a direct answer: the LHC could shed light on dark matter.

Dark matter

> "Is there any other point to which you would wish to draw my attention?"
> "To the curious incident of the dog in the night-time."
> "The dog did nothing in the night-time."
> "That was the curious incident," remarked Sherlock Holmes.
>
> Arthur Conan Doyle[4]

Like the dog that didn't bark at night in the Sherlock Holmes story, the universe contains matter that doesn't shine. This is a curious incident. The existence of matter that doesn't shine can be deduced from the gravitational pull exerted by invisible bodies. One of the most brilliant examples of this kind of deduction – worthy of a Holmes story – is the discovery of the planet Neptune. Urbain Le Verrier (1811–1877) understood that some unexpected distortions observed in Uranus's orbit could be imputed to the gravitational attraction of a still unknown planet. In England, John Adams had done similar, though less complete, calculations which were made public only later. But on 23 September 1846, Johann Galle of the Berlin Observatory received a letter from Le Verrier with the precise coordinates of a spot in the sky, at the border between the constellations of Capricorn and Aquarius, where he ought to make detailed observations. That very night, Galle identified a new celestial body, and observations during the next two nights confirmed that the newly discovered object was indeed a planet.

The discovery represented a triumph both for the Newtonian theory of gravity and for the power of human deduction. François Arago, speaking before the Académie des Sciences, poetically commented: "M. Le Verrier vit le nouvel astre au bout de sa plume" (Mr Le Verrier saw the new planet on the tip of his quill).

But not everything that doesn't bark is a dog. Mercury's orbit was showing some unexpected features too, with anomalies in the precession of its perihelion. So Le Verrier tried his luck again and proposed the existence of a new planet named Vulcan. But this time observations

---

[4] A. Conan Doyle, *Silver Blaze*, in *The Memoirs of Sherlock Holmes*, 1893.

did not confirm his hypothesis. As the discovery of Neptune had marked the triumph of Newtonian gravity, the perihelion precession of Mercury decreed its demise. The correct explanation of the anomaly in Mercury's orbit was to be found only in general relativity.

So the lesson is that an unexpected gravitational pull indicates either the presence of invisible mass or the need for modification of the theory of gravity. And this lesson, derived from the solar system, becomes very useful when we observe the universe at larger scales, where matter that doesn't shine is completely invisible to optical telescopes.

In the 1930s came the first hint for the presence in the universe of some element that exerted a gravitational pull, but was otherwise invisible. The Swiss astrophysicist Fritz Zwicky (1898–1974), studying a group of galaxies called the Coma cluster, observed that many of those galaxies appeared too fast to remain trapped inside the system. If you throw a rock up in the air, the rock will eventually come back, attracted by the gravitational pull of the earth. But if Superman throws a rock faster than 40 000 km/h, the rock will go into outer space and never come back. (Actually the atmospheric friction would disintegrate a rock with such high speed, but this is an irrelevant detail.) Zwicky found that many of the galaxies in the Coma cluster were like rocks thrown by Superman and should have quickly escaped from the group. Such fast galaxies could remain bound in the system only if the Coma cluster contained about 400 times more mass than visually observed. The idea of *dark matter* was born.

At that time, astronomical measurements of the velocities of galaxies were not precise enough to convince most astronomers that such a drastic hypothesis as the existence of dark matter was really necessary. But starting from the late 1960s, observational evidence in favour of dark matter grew, especially with the pioneering work by Vera Rubin and collaborators.

The rotational velocity of stars inside galaxies is an indicator of the amount of mass contained within the stellar orbit. Observations of spiral galaxies gave the unexpected result that much more mass ought to exist than was accounted for by visible stars. These results suggested that galaxies are like gigantic atoms with a nucleus made of stars in the centre of a uniform cloud of dark matter. This invisible cloud, much larger than the size of the visible central nucleus, is called the *galactic halo*.

Dark matter is ubiquitous in the universe and it is encountered in almost every large astronomical environment. An interesting way of detecting the presence of dark matter is based on the phenomenon called *gravitational lensing*. General relativity predicts that mass bends the trajectory of light, thus distorting images in a way similar to the effect of lenses in optics. Gravitational lensing has been used to measure mass in the universe through the visual distortion of background galaxies.

Figure 12.2 The visible part of the galaxy lies in the middle of a large cloud of dark matter, the galactic halo.

The pictures produced with this technique are very intriguing because they show real photographs of dark matter in action (see Figure 12.3).

The existence of dark matter can also be inferred from its role in the history of the universe. As we have seen, quantum fluctuations of the inflaton field provide the seeds for galaxies to form. After the period of inflation is completed, gravitational attraction makes these seeds grow, and the mass contained in the slightly denser regions of space contracts. But without dark matter the lifetime of the universe is not long enough to allow these seeds to grow into the dense galaxies and clusters of galaxies we observe today. Detailed numerical simulations show that dark matter successfully accounts for the formation of galaxies in the universe.

The empirical evidence in favour of dark matter is by now over-whelming. The combination of various astronomical observations with the measurements of the cosmic microwave background provides a precise determination of the amount of dark matter present today. It is found that, averaging over large distances, the universe contains *five times* more mass in dark matter than in ordinary matter. But the nature of dark matter is still a mystery.

Many possible explanations for the nature of dark matter have already been ruled out either by observations or by theoretical arguments. Every

Figure 12.3 A Hubble Space Telescope image of the galaxy cluster known as Abell 1689. The mass contained in the cluster in both forms of visible and dark matter acts as a gravitational lens, distorting the image of the galaxies located far behind.

Source: NASA / ACS Science Team / ESA.

attempt at attributing the evidence for dark matter to a modification of gravity has essentially failed. Thus, dark matter must correspond to some physical substance. One could imagine that dark matter is made of small planets or other celestial bodies that do not shine because they do not ignite the necessary thermonuclear reactions. But the following theoretical consideration excludes this possibility.

Calculations of nuclear reactions in the early universe predict the present density of the light chemical elements, like hydrogen, deuterium, helium, and lithium. The results of these calculations are in excellent agreement with the abundances of elements measured by astronomers and provide one of the most spectacular successes of the Big Bang theory. However, these results are very sensitive to the amount of protons and neutrons originally present in the universe, and are incompatible with the possibility that ordinary atomic matter is sufficiently abundant to account for dark matter. Thus dark matter cannot be made of planets or small stars. The study of gravitational lensing in our galaxy has given further support to this conclusion.

At this point, we have to refer again to Sherlock Holmes: "When you have eliminated the impossible, whatever remains, however improbable, must be the truth."[5] After eliminating the impossible, we are left with two solutions, both of which are quite surprising. The first solution has an astrophysical flavour: dark matter consists of a numerous population of black holes formed in the very early stages of the universe. We do not have a clear idea of how such a population of black holes could have been generated, but this possibility is not inconsistent with observations. The second solution has a particle-physics flavour: dark matter is made of some new kind of particles. Of course, from the point of view of the LHC, the second solution is much more interesting and that is the one I will pursue here.

We do not know yet what particle may constitute dark matter. However, many of the properties of this hypothetical particle can already be deduced from the known characteristics of dark matter. The dark-matter particle must be massive in order to account for the gravitational pull exerted by the invisible matter. It must be stable or else it would have disintegrated during the history of the universe and would not be present today. It must be electrically neutral and have no QCD charge so that it cannot bind with ordinary matter inside stellar material. The only known particle that satisfies these requirements is the neutrino, but unfortunately the neutrino is too light to make it a plausible candidate for dark matter. If a dark-matter particle really exists, it must be of a new and unknown kind.

The most intriguing aspect of the hypothesis that dark matter is made of a new kind of particle comes from a theoretical calculation. Let me explain the essence of this important result. When the universe was very hot and dense, particles of every sort were constantly produced and destroyed as a result of their frequent collisions. In an animal population some individuals die while new ones are born, keeping their total number approximately constant. In the same way at primordial times individual particles continually appeared and disappeared, but every particle species was roughly equally represented in the hot soup of the early universe. If a dark-matter particle really exists, it must have been an ingredient of this primordial soup and, at very early times, its population must have been as numerous as those of photons, electrons, or any other particle.

[5] A. Conan Doyle, *The Sign of the Four*, 1890.

As a result of its expansion, the universe cooled down. Once the universe cooled below a certain temperature, the particles inside the soup did not carry enough energy to create, in their collisions, the very massive dark-matter particles. At that stage, new dark-matter particles could no longer be produced. Just as a population of dinosaurs unable to procreate is destined to extinction, so the population of dark-matter particles in the universe could not be replenished and started to decrease. Dark-matter particles are believed to be stable and therefore they cannot spontaneously disintegrate, but they can disappear through a process called *annihilation*. Annihilation means that two dark-matter particles bump into each other and, in a mortal embrace, transfer their energy into other forms of particles and radiation, disappearing from the cosmos forever. However, the process of annihilation does not lead to extinction. As the universe keeps on expanding, it is increasingly difficult for the surviving dark-matter particles to find each other and annihilate. At a certain moment in the history of the universe, the mortal embraces of annihilation became so exceedingly rare that the population of dark-matter particles could no longer be reduced. At that stage, the number of dark-matter particles "froze-out", as is said in physics jargon, because neither could new particles be created nor old ones destroyed.

Theoretical physicists are able to compute the "frozen-out" remnants and determine the amount of dark-matter particles present in the universe today in terms of their physical properties. The result is that a new particle interacting through the weak force and with mass in the range between about 0.1 and 1 TeV has the right features to account for the observed density of dark matter. Such a particle is usually called a WIMP (for weakly interacting massive particle).

This result has caused much excitement in the community of particle physicists, because a value of mass between 0.1 and 1 TeV places the WIMP right in the middle of zeptospace territory. This is excellent news for the LHC, because it provides a new argument in favour of the existence of unknown particles in zeptospace. This connection between dark matter and particle physics provides a new motivation for exploring zeptospace that is completely independent of the arguments based on the problem of electroweak symmetry breaking or the naturalness problem. If the WIMP hypothesis is true, the LHC will discover that zeptospace is populated with a new form of matter, something five times more important than atoms in the mass content of the universe. The LHC might discover dark matter.

There is another reason why physicists find the idea of WIMPs so attractive. As described in Chapter 10, a fairly generic prediction of supersymmetry is the existence of a new massive, stable, and electrically neutral particle – the neutralino. This particle fits perfectly the role of WIMP. So the mysterious substance constituting dark matter could be nothing less than a shadow of superspace. It would be stunning to discover that supersymmetric particles are actually the most common form of matter in the universe, and that they are present all around us at this very moment.

Supersymmetry is not the only theory able to support the existence of WIMPs, and the identification of the dark-matter particle has become one of the most pressing experimental issues of today. Not only would it reveal the nature of this mysterious form of matter, but it could also expose a new and deeply important connection between cosmological history and the fundamental laws of particle physics.

Detecting dark matter

> To see what is in front of one's nose requires a constant struggle.
>
> George Orwell[6]

From the experimental point of view, a WIMP closely resembles a neutrino, although it is much more massive, probably being heavier than about a hundred protons. Just like neutrinos, WIMPs can also go through the earth almost unscathed because for them any material is essentially transparent. So the name "dark matter" is rather misleading. A dark body absorbs light, but does not emit it. WIMPs are not dark at all; they are "invisible" because they are perfectly transparent. But I suppose that "transparent matter" sounds a lot less mysterious than "dark matter".

If WIMPs are produced in proton collisions at the LHC, they cannot be directly detected because they interact too weakly to leave any trace in the instruments. However, their presence can be inferred from measurements of "missing energy". The experimental signal of "missing energy" was described in Chapter 10 in the case of supersymmetry, which indeed provides the prototype example of a WIMP. Although the discovery of "missing energy" can be regarded as indicative of the artificial production of dark matter at the LHC, it cannot be considered as a conclusive proof. Other kinds of invisible particles, unrelated to dark matter, could produce the same experimental result. The definitive evidence for the discovery of dark matter requires the confirmation that the new particles produced at the LHC are the same as those that constitute the galactic halo. Such evidence cannot come from collider experiments.

Our galaxy, the Milky Way, like any other galaxy in the universe is embedded in a large halo of dark matter. If WIMPs really constitute

---

[6] G. Orwell, *In Front of Your Nose*, Tribune, 22 March 1946; reprinted in *The Collected Essays, Journalism and Letters of George Orwell*, Harcourt, New York 1968.

dark matter, these particles are physically present all around us, not only in the sky but also on earth. Every litre of air we breathe contains several of these galactic WIMPs. But WIMPs do not remain stuck in our lungs. On the contrary, they move around space very fast. Every second, hundreds of millions of galactic WIMPs go through your body at the speed of about one million km/h. And yet they leave almost no trace because any material is almost perfectly transparent to them.

Many ongoing experiments are trying to detect the presence of WIMPs, while new and more sensitive instruments are planned for the future. The direct detection of galactic WIMPs represents a formidable experimental challenge for at least two reasons. First of all, the density of WIMPs in our surroundings is very low. In every cubic kilometre of space around us there is half a billionth of a gram of dark matter; this is equivalent to half a kilogram of dark matter in the space occupied by the whole earth. Only when we average over large portions of the universe do we find that there is more mass in dark matter than in ordinary matter. But we happen to live on a planet that presents an unusual concentration of atoms and molecules with respect to an average place in the universe. So, on earth, dark matter is relatively rare.

The second challenge for experimental detection is that WIMPs interact very weakly, as their name suggests. Occasionally, in their frenzied motion at one million km/h, WIMPs can bump into an atomic nucleus, depositing some energy into the material during the collision. The problem is that the power generated by dark-matter collisions in a kilogram of material is about 10^{-19} watts. This is a really minute quantity. To extract an experimental signal as intense as that emitted by a normal 100-watt light bulb, one would need to build a detector weighing 10^{18} tonnes – about 1 per cent of the mass of the moon. But physicists cannot use large pieces of the moon as detectors and must instead devise clever ways of revealing the minute energy deposits of WIMPs. It is absolutely stupefying that experiments can achieve the sensitivity required to detect such an extremely feeble signal.

The first requisite for an experiment aiming to detect the presence of galactic dark-matter particles is the shielding of the apparatus from any source of energy that could mask the signal sought for. An appropriate screening of cosmic rays demands that the experiments take place under layers of rock several kilometres thick. For this reason, such experiments are located in underground laboratories built either in inactive mines, like the Soudan Laboratory in Minnesota, or off motorway tunnels under mountains, like the Gran Sasso Laboratory in Italy. The experimental detectors must also be shielded from the natural radioactivity of the rock, which is much more intense than the dark-matter signal.

When a WIMP hits a nucleus of the detector material, it kicks it off from its position inside the crystalline structure. In the collision, the nucleus acquires some energy, which is then released either in the form of ionisation, or as acoustic waves produced by vibrations inside the crystal lattice, or as an increase of temperature in the material. Experiments use different methods to detect these small energy deposits, but all of them employ extraordinary technologies. For instance, some experiments operate at temperatures only a few thousandths of a degree above absolute zero. At such low temperatures, the special material used for the detectors reacts to the deposition of even infinitesimal amounts of heat energy with a significant change in temperature, which can then be measured.

A different technique for revealing the existence of dark-matter particles exploits the fact that WIMPs can still occasionally annihilate in the universe today. When two WIMPs mutually kill each other in the annihilation process, they transform their energy into other kinds of particles, which can be detected by our instruments. The death of WIMPs can produce gamma rays, neutrinos, positrons, electrons, anti-protons, and antideuterium nuclei, which can be identified by satellite or earth-based detectors, after subtracting the effect of the ever-present flux of cosmic rays. Neutrinos are especially suited for the identification of the death of WIMPs, but they are very elusive particles and their observation requires detectors distributed over large areas. One of these detectors consists of a grid of photomultiplier tubes arranged in various strings at a depth of about 2500 m under the Mediterranean Sea. Another experiment makes use of strings drilled inside the ice of Antarctica at depths between 1500 and 2500 m, taking advantage of the fact that the ice in the South Pole is extremely transparent about one kilometre below the surface.

So far, no experiment searching for galactic dark-matter particles has unequivocally established the existence of WIMPs, although there are several still controversial claims of detected signals. More data is necessary before drawing definite conclusions, and the LHC will add crucial information.

Just as a police inspector needs to gather many clues before being convinced of the guilt of a suspect, so the identification of dark matter will require corroboration from different experiments. The various experimental strategies for the discovery of WIMPs are complementary, for they give independent information about the nature of the dark-matter particles. The artificial production of dark matter at the LHC will allow the best measurements of the intrinsic properties of WIMPs, such as their mass and interaction strength. The direct detection of dark matter in underground experiments will prove that WIMPs are physically present around us and will provide us with a measurement of their density. The observation of the death of WIMPs will give us information about their distribution in our galaxy. Only the comparison of the results from these different experimental techniques can give conclusive evidence for the existence of dark-matter particles in the universe.

Dark energy

Obscurum per obscurius, ignotum per ignotius.
[Dark through the darker, unknown through the more unknown.]

Alchemy motto

Supernovae are one of the most dramatic and violent phenomena that we can observe in nature. When certain aging massive stars exhaust their nuclear fuel, the gravitational force causes the star to collapse under its own weight. This produces an enormous explosion in which stellar material is ejected with speeds up to a hundred million km/h. A supernova can look as bright as billions of suns, outshining its entire galaxy. In a period of a few weeks, a supernova can emit as much energy as our sun during its entire life span.

The so-called type Ia supernovae are very useful to map the distant universe not only because they are so luminous, but also because they burn with the same intrinsic brightness. This property offers a reliable method to measure their distances. Just as a light bulb looks dimmer the further away it is, the observed brightness of a type Ia supernova is a measure of its distance from us.

Distant supernovae move away from us as a result of the expansion of the universe. Their recession speeds can be determined by measuring the wavelength of the light they emit. Just as we hear the siren of an ambulance wailing at a lower pitch as it drives away from us, so the wavelength of the light emitted from a receding astronomical body is longer, in proportion to its velocity. The determination of how fast distant objects move away from us carries the information of how fast the universe is expanding. So observations of type Ia supernovae provide a way of measuring the expansion rate of the universe.

Inflation is the initial engine that set into motion the expansion that we observe today. But after inflation was over, the attractive gravitational pull from all the matter and radiation present in the universe contributed to slowing down the expansion rate. Therefore today the expansion of the universe should be *decelerating*. In the 1990s several groups of astronomers started a programme of observations of distant supernovae to measure the rate of deceleration in the expansion of the universe. The project was very difficult because, although supernovae have distinctive features easily identifiable, nobody can predict where or when a supernova will occur. Moreover, supernovae are rare phenomena, as they happen in a galaxy approximately only once every hundred years or so. The programme of observations was possible only because of special technologies applied to astronomical instruments capable of simultaneously monitoring large numbers of galaxies.

In 1998 two groups of astronomers announced their results. It was one of the most unexpected discoveries in science in many years: the expansion of the universe is not decelerating, but instead is *accelerating*. This result is quite shocking. If the expansion of the universe is speeding up, it means that some form of antigravity exerting a repulsive push must be present today. An invisible engine is powering the expansion of the universe at this very moment. Combining observations on supernovae and other astronomical measurements with the most recent data on the cosmic microwave background, it is found that 72 per cent of the energy present in the universe today is in the form of a new substance exerting repulsive gravity. This mysterious substance has been named *dark energy*.

It was disconcerting enough to learn that dark matter is five times more important than ordinary matter in the mass content of our universe, but now we are confronted with the startling fact that dark matter is dwarfed by an even more uncanny substance. Dark energy counts in the energy balance of our universe almost 16 times more than ordinary atomic matter. This result has shaken the particle physicists' beliefs in their understanding of the world, but at the same time has given them new reasons to look beyond their present theories. The Sublime Marvel (also known as the Standard Model) of particle physics can explain only 4.6 per cent of the ingredients of the universe; the rest is just food for theoretical speculation.

We have no compelling idea what might be the origin of dark energy. Solely from its repulsive gravitational push, we deduce that dark energy must be a form of cosmological constant or something resembling it very closely. By normal standards, the energy content of dark energy in our neighbourhood is not particularly impressive. The dark energy stored in a kilometre cube of space corresponds to the energy used by a 60-watt light bulb in one hundredth of a second. This should discourage even a science-fiction writer from imagining that dark energy could be used by humanity (or aliens) as an energy source. But dark energy is believed to uniformly fill all space and thus its total amount is enormous.

Dark energy is by far the most common form of energy present in our universe today. But this was not always the case during cosmological history. While matter and radiation are diluted by the expansion of the universe, the cosmological constant (as the name suggests) fills space always in the same way. Therefore, if dark energy consists merely of a cosmological constant, it had a negligible effect in the early universe relative to other forms of energy, but its role grew more important with time. The cosmological constant is destined to become essentially the only form of energy present in the universe. In the future, the expansion

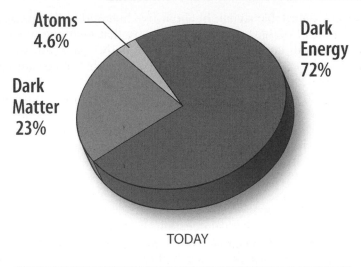

TODAY

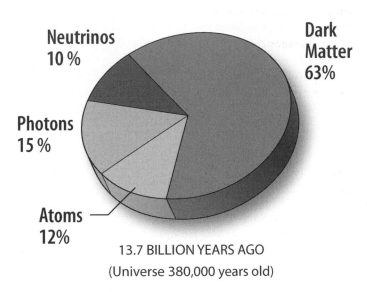

13.7 BILLION YEARS AGO
(Universe 380,000 years old)

Figure 12.4 The energy content of the universe today (top frame) and 13.7 billion years ago (bottom frame), when the cosmic microwave background was emitted and the universe was only 380 000 years old.

Source: NASA / WMAP Science Team.

of the universe will grow ever faster, incessantly accelerated by the invisible engine of dark energy. Far regions of space will be pushed away from us at a rate exceeding the speed of light, disappearing forever from our view. In about a hundred billion years from now, anything beyond Andromeda will be receding so fast that it will not be able to send us any detectable signal. In about five hundred billion years from now, space will be blown apart so violently that anything beyond the solar system will be forever invisible. Eventually, space will stretch so rapidly that a man will no longer be able to see his feet (assuming that humanity still existed).

One should not be surprised that the expansion of the universe can proceed at rates faster than light. Special relativity teaches us that no signal can travel through space at a speed faster than light. But nothing forbids space itself to expand at a rate faster than light, as long as no information is transmitted between two physical points at that speed. Superluminal expansion was commonplace almost everywhere at the time of inflation.

We still do not know whether our universe will meet the dismal fate predicted by the cosmological constant. Dark energy could be a form of cosmological constant that varies with time and its effect could diminish or disappear in the future, before becoming as brutal as previously described. Just as the primordial inflationary period ended when the inflaton field underwent a phase transition, so dark energy could also eventually disappear from space. The actual destiny of the universe will not be determined until we discover the true nature of dark energy.

Stepping into the multiverse

> There is no law except the law that there is no law.
> John Archibald Wheeler[7]

The discovery of dark energy puzzled cosmologists, who did not expect a cosmological constant to exist, let alone one with such a large effect. Particle physicists were puzzled too, but essentially for the opposite reason. They cannot understand why the cosmological constant is so small. The problem with the cosmological constant is completely analogous to the naturalness problem of the Higgs substance encountered in Chapter 9. The quantum fluctuations of virtual particles in the vacuum affect various physical quantities and, in particular, lead to the expectation that the cosmological constant should be 10^{120} times larger than is actually observed. This result is known as the worst-ever theoretical prediction in physics. It is as off the mark as predicting that the volume of a proton must be roughly equal to the volume of the observable universe.

---

[7] J.A. Wheeler, as quoted in J.D. Barrow and F.J. Tipler, *The Anthropic Cosmological Principle*, Oxford University Press, Oxford 1986.

The naturalness problem of the Higgs substance has been a fruitful source for many new theoretical ideas, such as supersymmetry, extra dimensions, technicolour, and several others. On the other hand, every attempt to solve the naturalness problem of the cosmological constant has failed. Is this an indication that the naturalness problem is based on a false prejudice and that physicists are following the wrong trail?

The crisis with the concepts of dark energy and the cosmological constant has stimulated new approaches to some of the basic questions about particle physics. One interesting possibility has emerged as a consequence of a feature of string theory, called the *landscape*. The physicist Leonard Susskind took this term from biochemistry, where it refers to the vast number of possible configurations of large biomolecules. String theory too allows for a huge number of possible configurations, each of them corresponding to a different particle world, each with its own characteristics and its own values of the fundamental constants.

An interesting result arises when the landscape is married with the idea of *eternal inflation*. In eternal inflation the universe is perpetually generating bubbles of space, which in turn expand and form other bubbles in their interiors. Each bubble corresponds to a different configuration of the landscape. So the entire cosmos ends up divided into a huge number of separated universes, each of which corresponds to a different particle world. For this reason, this scheme is called the *multiverse*, in contrast with the usual case of a single *universe*.

The multiverse is simultaneously populated with a huge variety of universes, each of them characterized by different physical laws. This conception of reality defies the idea that a single theory, uniquely determined by logical consistency, must lie on top of Jacob's ladder. The idea of the multiverse forces us to reconsider some of the fundamental "why" questions in physics and to wonder whether they are good leads or just red herrings. According to this new conception, the physical laws that we observe in nature are only one of the many possible alternatives that simultaneously coexist in the multiverse. Let me illustrate these ideas with an example.

In 1595 Johannes Kepler asked himself the question: "Why are there six planets?" This sounds like a good "why" scientific question, because the six planets (Uranus and Neptune had not yet been discovered at that time) were believed to be fundamental elements of the universe. This question is similar to those that we ask today, like: "Why are there three generations of quarks and leptons?" In his *Mysterium Cosmographicum*, Kepler proposed an answer to his question based on geometrical symmetry. Planetary orbits, claimed Kepler, lie on successive spheres that circumscribe and inscribe the five Platonic solids. The Platonic shapes are convex solids with identical faces and were believed, in ancient times, to have magical properties. Since there exist only five

Platonic solids (regular pyramid, cube, octahedron, dodecahedron, icosahedron), there must exist only six planets. The power of Kepler's hypothesis was its capability of predicting the ratios of planetary distances, which matched fairly well the observations of the time. Later in life, Kepler pursued this line of thought in a more mystical vein and suggested in his *Harmonices Mundi* that planetary positions and orbital velocities, together with all proportions of the natural world, follow the rules of musical harmonics. The *musica universalis*, resembling the Pythagorean concept of "music of the spheres", drives the motion of the cosmos in a heavenly symphony. Arcane as it may seem, this hypothesis led Kepler to formulate his third law of planetary motion.

Of course today we know that Kepler's geometrical construction of the solar system is wrong. Kepler's original "why" question was not the trail towards a fundamental problem. There is nothing fundamental in the number of planets or in their distances from the sun. Our solar system is just one out of the many that exist in the universe, each of them having different numbers of planets and different planetary distances. From a terrestrial point of view, we may be surprised at the peculiar coincidence that the distance between the earth and the sun is just right to allow for the existence of liquid water. It is indeed a lucky circumstance that we do not have to roast in the Mercurial heat or freeze in the Uranian cold. But, from a broader perspective, there is nothing mysterious in this fact. Out of the many available planets in the universe, our form of life developed only where water *could* exist in the liquid state. So there is little surprise that we happen to live on a planet at just the right distance from the sun.

The multiverse extends these considerations to a much larger scale. Out of the many available universes, we must necessarily find ourselves in one of those that have the right features to support life. Some of the properties of our universe can be deduced from the mere fact that such a universe exists, just as the earth–sun distance can be approximately inferred from the requirement that liquid water must exist on our planet. Following this kind of reasoning, Steven Weinberg found that only universes with a sufficiently small cosmological constant are hospitable to life. If the cosmological constant were much larger, its repulsive gravity would have blown the galaxies apart and no place in the universe would be inhabitable. Maybe the observed dark energy just reflects a generic property of the part of the multiverse where humans can exist and contemplate the cosmos. In this context, the concept of naturalness is ill-posed, because what superficially appears as a contrived fine-tuning could be just the consequence of the narrow part of the multiverse available to human existence.

A fraction of the scientific community is vehemently against the idea of the multiverse. Some physicists view the use of the condition that life must develop in our universe too anthropocentric. But actually the multiverse represents a complete denial of anthropocentrism. Not only is the earth a generic planet in the cosmos, but even our universe is just a generic element of the multiverse. If true, this would be the ultimate Copernican revolution.

Another criticism arises from the belief that the multiverse represents a defeat of the human desire for deduction and a retreat from the primary goals of scientific investigation. But, while the multiverse certainly requires a profound reformulation of some of our current scientific questions, it does not necessarily imply complete chaos in the physical laws. At any rate, physics is a natural science and this issue will not be resolved by philosophical prejudices. The concept of the naturalness problem in the Higgs substance will be experimentally tested and the result will give us some indications in favour or against some of the ideas based on the multiverse. So in this case too, the LHC goes straight to the heart of the matter.

13
Epilogue

━━━━━◦◦◦◦━━━━━

It is bad enough to know the past; it would be intolerable to know the future.

William Somerset Maugham[1]

The LHC is a fantastic intellectual adventure. The goal of this adventure is not just to discover some new particles or to comprehend the functioning of a world remote from our sensory perception. Its goal is much more universal. The LHC represents one further step in the long path undertaken by humanity to understand the meaning of physical phenomena, the structure of matter, the principles of nature, and the fundamental laws that govern the universe. It is an essential part of the global human endeavour that we call science. The questions addressed by the LHC project, though masked by the mathematical complexity of the theories of particle physics and the technological intricacy of the experimental apparatus, are the same that have puzzled human intellect since the dawn of civilization.

But the effects of the LHC project on society have ramifications well beyond its primary goal of scientific discovery. The planning, development, and construction of the LHC required advances in many of the frontiers of technology. Such a rapid growth of innovation can only be attained in large and complex scientific projects devoted to fundamental research, because the activity is too risky and uncertain for private industry. These kinds of scientific projects accelerate technological progress in a way that would be otherwise impossible and unimaginable for society.

The LHC teams were constantly confronted with the need for research on new materials and on new instrumentation, with the necessity of developing innovative electronics, and with the problem of dealing with formidable amounts of digital information. In their everyday work,

[1] W. Somerset Maugham, as quoted in R. Hughes, *Foreign Devil*, Deutsch, London 1972.

the scientists were continually challenging what is possible. The LHC offered a unique opportunity to bring together the financial resources and the human intellect necessary to meet these challenges and break new ground in technological inventions. Science channels human talent and creativity towards solving complex problems whose solutions invariably lead to unexpected applications. Rarely can we predict what these applications will be, but they unfailingly arrive.

Research on particle accelerators and detectors keeps on producing rich harvests of spin-offs useful for society, ranging from hadron cancer therapy to synchrotron radiation, from positron emission tomography to magnetic resonance imaging, and other tools for medical diagnosis and imaging technology. The extreme requirements of particle-physics experiments in information technology have always been the drive for innovation, a prime example being the World Wide Web invented at CERN. Today the LHC provides an entirely new challenge, generating one million gigabytes of data per second, if all events were recorded. These data must be quickly sifted through to select ten million gigabytes per year, subsequently stored and analysed in different parts of the world. These stringent requirements provide an ideal opportunity to develop and test new computing and information technologies, like the GRID, that may one day become part of our everyday life.

The large scale of a project the size of the LHC requires direct involvement with industry. This leads to immediate economic benefits for the industrial sector, but also to indirect spin-offs such as the development of new manufacturing techniques and the acquisition of new skills and expertise. Moreover, the international character of the LHC has favourable political implications in the relations and the cooperation between different countries. Science promotes exchange and understanding, bringing together nations, institutions, and individuals.

Last, but not least, the LHC provides a unique education and training opportunity for students and young scientists. Young people are instrumental in the activities of the LHC and are often the driving forces behind many specific tasks. These people learn the ability of addressing complicated problems, of mastering advanced technologies, of adapting to challenging situations, and of working in large teams. Not all of these people remain in the field of scientific research, and they often carry their unique skills and experience into other sectors of society. Investments in the LHC are also investments in future generations of capable and competent individuals.

But there is little doubt in the minds of the people who are working at the LHC that discovery is the primary goal. As passionately expressed by the mathematician and theoretical physicist Henri Poincaré: "The scientist does not study nature because it is useful to do so. He studies it because he takes pleasure in it, and he takes pleasure in it because it is beautiful. If nature were not beautiful it would not be worth knowing, and life would not be worth living."[2]

How will the zeptospace odyssey end? Homer narrates that Ulysses concluded his wanderings around the Mediterranean by finding his way back to Ithaca. After slaughtering all the suitors of Penelope, Ulysses regained his place as King of Ithaca, and finally brought some peace to the island. So the Odyssey ends. The poet Dante Alighieri did not know any Greek and apparently never read the Odyssey, save for some medieval summaries and the references made in Ovid's Metamorphoses. In his blissful ignorance, Dante invented an interesting addition to Homer's epic poem. Dante imagined that, once back in Ithaca, Ulysses could not restrain his urge for further explorations and, after some years, he set off again on a new journey with his most faithful companions. He exhorted them by saying that "you were not born to live like mindless brutes but to follow paths of excellence and knowledge."[3] So they sailed westwards, pushed by their desire to gain knowledge of the unknown, and crossed the Pillars of Hercules, beyond which no man was permitted to venture. "And with our stern turned toward the morning light, we made our oars our wings for that mad flight."[4]

We can find inspiration from these stories and try to imagine the end of the zeptospace odyssey. The discovery of the Higgs boson can be likened to the successful return of Ulysses to Ithaca, as narrated by Homer. This discovery will be the confirmation that the idea of spontaneous breaking of the electroweak symmetry is correct and it will complete the experimental validation of the Standard Model by detecting its final missing ingredient. There is no doubt that finding the Higgs boson will be a crucial step in the understanding of the principles of nature and in our knowledge of the particle world.

Nevertheless, to many physicists, the Higgs boson will seem an anticipated discovery. We have already gathered experimental data and theoretical clues that point in the direction of the Higgs boson, giving us a certain confidence to believe that this particle exists. The situation is somewhat similar to the case of the discoveries of the W and Z at CERN in 1983 and of the top quark at Fermilab in 1995. Even before these particles were discovered, theory was providing strong indications for their existence. But unlike the W, Z, or top quark, for which theory had precisely anticipated their properties, the predictions regarding the Higgs boson today are a lot less sharp. The part of the Standard Model

---

[2] J.H. Poincaré, *Science et Method*, Flammarion, Paris 1908.

[3] D. Alighieri, *The Divine Comedy, Inferno*, Canto XXVI, 119–120, translated by M. Musa.

[4] D. Alighieri, *The Divine Comedy, Inferno*, Canto XXVI, 124–125, translated by M. Musa.

associated with the Higgs boson is determined only by a choice of simplicity and not dictated by any profound principle. This simple choice may well turn out to be wrong. Nature may have good reasons for selecting a different structure responsible for breaking the electroweak symmetry, or perhaps the Higgs boson is the remnant of still unknown forces acting in zeptospace. The experimental detection of the Higgs boson will then hold for us surprises, since the properties of the new particle may turn out to be quite different from those expected in the Standard Model.

But just as Dante added an inventive finale to Homer's epic, so the zeptospace odyssey may not simply end with the discovery of the Higgs boson. Physicists eagerly hope for and expect a new twist of the story. And this hope is not just based on some generic pretence that the LHC explores an unknown and uncharted territory. On the contrary, the unsatisfactory features related to the Higgs boson, the naturalness problem, and the dark-matter connection provide good arguments for believing that zeptospace is populated by phenomena and particles other than just a simple Higgs boson.

An extraordinary result of 19th-century science was showing that celestial bodies are made of the same chemical elements present on the earth, demonstrating that the whole universe is composed of the same kind of matter. Recent cosmological observations have shaken this picture. Ordinary atomic matter constitutes less than 5 per cent of the content of the universe, while the rest is in the form of still unexplained and unknown substances: dark energy and dark matter. The LHC has the chance of revealing the identity of dark matter, solving one of the biggest puzzles about our present universe.

One of the main themes in theoretical physics of the past decades has been the study of what zeptospace might look like. In this vein, many new theoretical ideas have been proposed. Some of these ideas undoubtedly appear so contrived and complicated that they introduce more problems than they provide solutions. Others have failed when confronted with experimental data from previous colliders, especially LEP. But the process of speculative investigation of zeptospace has produced some fascinating new ideas, like supersymmetry, extra dimensions, technicolour, and several others. These ideas entail a complete revision of our conception of space-time, forces, and symmetries. Their experimental discovery would generate a revolution of our views of physical reality with intellectual consequences comparable to those of relativity or quantum mechanics.

Recent studies have revealed deep and unexpected connections between various theoretical ideas about zeptospace, showing that the different proposals are not mutually exclusive and that nature could have simultaneously employed several of them in shaping zeptospace. Moreover, once a new idea is generated, it tends to take on a life of its own, like a genie out of a magic lamp. As Victor Hugo put it: "Ideas can no more flow backward than can a river."[5] New ideas sometimes lead to results never imagined by their originators. Einstein introduced the cosmological constant to make the universe static; little did he imagine the explosive phenomenon of inflation or the dark energy that dominates the universe today. Yang and Mills developed their theory in a failed attempt to model pion interactions, but they could not have suspected that gauge theories contain the fundamental principle that governs the particle world. String theory was invented to describe hadrons before the advent of QCD, and today is considered the most credible candidate for unification of gravity with the other known forces. Science does not always advance along a straight and logical path, but follows unexpected routes. Some of the ingenious theories invented for zeptospace may have nothing to do with the discoveries made at the LHC, but one day we may find that they play a crucial role in a completely different scientific context. This will be yet another spin-off of the LHC project, a spin-off relating to conceptual and intellectual developments.

The situation regarding the theoretical predictions of zeptospace is still very unclear. Each of the proposals has some interesting features but also some drawbacks, and not one of them has yet emerged as the most probable theory. This state of uncertainty feeds excitement into the experiments at the LHC. Many puzzles and problems still haunt zeptospace, and physicists thrive on puzzles and problems. The search for the physics of zeptospace is very much like the exploration of an unknown territory, where precious treasures are believed to exist but no one has a map that describes where to find them. We do not know what lies in zeptospace and the LHC has just started its journey.

[5] V. Hugo, *Les Misérables*, Pagnerre, Paris 1862.

Acknowledgements

This book grew out of public lectures and seminars I have given on the subject, out of the wish to communicate to people outside the world of physics the significance of the approaching discoveries, and out of the desire to impart to them some of the awe and excitement I personally feel at being part of this historic project. Though expanded in scope, the text maintains the original style of an oral presentation. As such, its aim is limited to inviting the general reader to share one physicist's enthusiasm and expectations of the LHC results, and hence there is no claim here of completeness.

I want to thank all the people at CERN for the many conversations I have had with them on topics related to the LHC, for their eagerness to discuss physics, and for making this laboratory the most wonderful and most stimulating working environment that I have ever had the pleasure of being a part of. I am especially grateful to Guido Altarelli, David Barney, Fabiola Gianotti, Thomas Lohse, Ken Peach, Antonio Riotto, Gigi Rolandi, Emma Sanders, and James Wells for their careful and thorough reading of the manuscript and their important comments which have much improved this book. I have profited greatly from many discussions with Reyes Alemany, Luis Alvarez-Gaumé, Ignatios Antoniadis, Tiziano Camporesi, Albert De Roeck, Alvaro De Rújula, Gia Dvali, John Ellis, Sergio Ferrara, Christophe Grojean, Karl Jakobs, Julien Lesgourgues, Michelangelo Mangano, Emilio Picasso, Riccardo Rattazzi, Lucio Rossi, Geraldine Servant, Mike Seymour, Elena Shaposhnikova, Ezio Todesco, Joachim Tuckmantel, Gabriele Veneziano, Jörg Wenninger, and Urs Wiedemann. I also wish to thank Mariarosa Mancuso and Nicoletta Moncada for comments on an earlier version.

I have greatly benefited from the excellent book collection of the CERN library, and I am grateful to its staff for their kindness and skill in finding documents for me. I am particularly thankful to Tullio Basaglia for his generous help, to Anita Hollier of the Pauli Archive, and to Christine Sutton, editor of the CERN Courier.

The staff at Oxford University Press has made this project a pleasure for me with their professional, yet friendly, guidance. Sonke Adlung, Melanie Johnstone, and April Warman have been particularly helpful.

I wish to thank Paul Beverley for his valuable editing assistance and for carefully improving my English.

My wife Debra has read more than one version of the manuscript and has made a number of valuable comments. The questions of my sons Giacomo and Enrico have stimulated me in finding ways to present many of the advanced physics concepts. I dedicate this book to my family.

Glossary

Accelerator   A machine that accelerates beams of particles to high energies.

ALICE   (A Large Ion Collider Experiment) The LHC detector devoted especially to the study of heavy-ion collisions.

Alpha particle   A helium nucleus, consisting of two protons and two neutrons.

Annihilation   The process in which a particle and an antiparticle disappear, transforming their energies into other forms of particles and radiation.

Antimatter   Matter made of antiparticles.

Antineutron   The antiparticle of the neutron.

Antiparticle   The combination of quantum mechanics with special relativity predicts that each particle has a corresponding antiparticle. The antiparticle has the same mass and spin of the particle, but opposite electric charge.

Antiproton   The antiparticle of the proton.

Asymptotic freedom   The property of a fundamental force to become arbitrarily weak when probed at shorter distances.

ATLAS   (A Toroidal Lhc ApparatuS) One of the detectors that studies the collisions at the LHC.

Atom   A building block of matter, consisting of a positively charged nucleus surrounded by a cloud of electrons.

Atomic number   The electric charge of a nucleus in units of the proton charge, thus equal to the number of protons in a nucleus.

Atomic weight   The weight of a nucleus in units of 1/12 of the mass of carbon-12, thus approximately equal to the number of protons and neutrons in a nucleus.

Background   The expected rate of events produced in the collisions, according to simulations based on the Standard Model.

Baryon   Any composite particle (such as the proton and the neutron) made of three quarks held together by gluons.

BNL   (Brookhaven National Laboratory) The US research laboratory established in 1947 and located in Upton, New York.

Boson   Any particle with zero or integer spin.

Brane   An entity with fewer dimensions than the space in which it is embedded, where particles and forces can be confined.

CERN   (Conseil Européen pour la Recherche Nucléaire) The European laboratory for high-energy physics established in 1954 and located near Geneva, Switzerland.

Chargino   An electrically charged hypothetical particle predicted by super-symmetry.

Classical physics   The set of physical laws that do not include quantum mechanics and relativity.

CMS (Compact Muon Solenoid) One of the detectors that studies the collisions at the LHC.

COBE (COsmic Background Explorer) The NASA satellite, launched in 1989, that first identified the temperature fluctuations in the cosmic microwave background.

Collider An accelerator with two counter-rotating particle beams that collide at certain designated points.

Colour The particle charge associated with the strong force as described by QCD (analogous to the electric charge of QED).

Compactification The process in which some spatial dimensions curl up in small regions of space.

Cosmic microwave background The electromagnetic radiation filling the universe, which peaks in the microwave range and has a thermal spectrum with a temperature of 2.7 degrees above absolute zero.

Cosmic rays High-energy particles of astrophysical and cosmological origin that continuously strike the earth.

Cosmological constant A uniform distribution of energy that does not correspond to any ordinary form of matter.

Cosmology The study of the evolution of the universe.

Coupling constant A number defining the strength of a force.

Cryogenic system The distribution system of liquid helium used to cool the LHC dipoles to 1.9 degrees above absolute zero.

Dark energy A still unknown form of energy that exerts negative pressure, constituting 72 per cent of the energy content of the universe.

Dark matter A still unknown form of non-luminous matter constituting 23 per cent of the energy content of the universe.

Decay The process in which a particle disappears, transforming its energy into other forms of particles and radiation.

DESY (Deutsches Elektronen SYnchrotron) The German laboratory for fundamental research established in 1959, with sites in Hamburg and Zeuthen.

Detector The assembly of instruments used to measure the particles produced in the collisions between the two proton beams.

Dipole magnet A device producing a magnetic field used to steer the proton beams into circular trajectories.

Effective field theory A quantum field theory valid within a certain range of energies, obtained by truncating the effects of small distances.

Eightfold Way A method of classifying hadrons that reveals their symmetry properties.

Electromagnetic calorimeter The instrument for measuring the amount of energy carried by electrons and photons.

Electromagnetic force One of the four fundamental forces. It is responsible for all electric and magnetic phenomena.

Electron An elementary particle with negative electric charge, which is a constituent of atoms.

Electronvolt (eV) A unit of energy or mass, corresponding to the energy gained by an electron when accelerated in vacuum by an electrostatic potential of one volt.

Electroweak symmetry breaking The still unknown process that generates masses for quarks, leptons, and gauge particles. The Higgs mechanism seems to be the most likely source of electroweak symmetry breaking in nature.

Electroweak theory The theory that describes both electromagnetic and weak forces in a single conceptual structure.

End-cap   The part of the detector placed at each of its two ends used to provide coverage of particles moving relatively close to the proton beams.

Event   The set of all particles produced by the collision between two energetic protons at the LHC.

Fermilab   (Fermi National Accelerator Laboratory) The US laboratory for research in particle physics established in 1967 and located in Batavia, Illinois.

Fermion   Any particle with spin equal to ½ or to an odd multiple of this quantity.

Galactic halo   The galactic component that extends far beyond the visible part of the galaxy and that exerts a measurable gravitational pull.

Gauge particle   A particle that communicates a fundamental force.

Gauge theory   A quantum field theory based on a symmetry principle, which describes a fundamental force.

General relativity   The theory that describes gravity in terms of the curvature of space and time.

Generation   Each of the three sets of quarks and leptons present in the Standard Model.

GeV   A unit of energy equal to one billion electronvolts.

Gluino   A hypothetical particle, which is the supersymmetric partner of the gluon.

Gluon   The particle that communicates the strong force.

Grand unified theory   A hypothetical theory in which electromagnetic, weak, and strong forces merge into a single force at very small distances.

Gravitational force   One of the four fundamental forces, acting on any form of mass and energy.

GRID   A distributed computer network sharing computing power and data storage.

Hadron   Any composite particle made of quarks, antiquarks, and gluons.

Hadronic calorimeter   The instrument for measuring the amount of energy carried by hadrons.

HERA   (Hadron Elektron Ring Anlage) The accelerator colliding a beam of protons with a beam of electrons or positrons that operated at DESY between 1992 and 2007.

Hierarchy   The largeness of the ratio between the weak and the Planck lengths (equal to 10^{17}), expressing the feebleness of gravity with respect to the other forces.

Higgs boson   The new particle associated with the Higgs mechanism.

Higgs mechanism   The spontaneous breaking of the electroweak symmetry induced by a uniform distribution of a quantum field, called the Higgs field. This mechanism can generate masses for the elementary particles of the Standard Model.

Higgs substance (or Higgs vacuum expectation)   The space-filling uniform distribution of the Higgs field at the origin of the Higgs mechanism.

Inflation   The theorized initial rapid expansion of the universe, which produced the conditions capable of explaining the present structure of our cosmos.

Inflaton field   The hypothetical quantum field that triggered inflation during the early stages of the universe.

Ion   An atom that has acquired or lost some electrons, thus carrying a net electric charge.

Jet   The stream of clustered hadrons that surrounds a quark or a gluon emerging from a collision.

Kaluza–Klein mode   One of the series of particles with increasing masses, which are the manifestation of a particle moving in a space with extra dimensions.

Landscape   The huge set of theories of the particle world, each characterized by different physical laws, that could emerge from string theory.

LEP   (Large Electron–Positron collider) The accelerator colliding electron and positron beams that operated at CERN from 1989 to 2000.

Lepton   The class of particles that are not affected by the strong force, consisting of the electron, the muon, the tau, and the three neutrinos.

LHC   (Large Hadron Collider) Read this book and find out.

LHCb   (Large Hadron Collider beauty experiment) The LHC detector devoted especially to the study of hadrons containing bottom (or beauty) quarks or antiquarks.

LHCf   (Large Hadron Collider forward) The LHC detector devoted to the study of simulated cosmic rays in laboratory conditions.

Luminosity   The number of particles per square centimetre per second accelerated in the high-energy beam.

Meson   Any composite particle (such as the pion) made of a quark and an antiquark held together by gluons.

Missing (transverse) energy   An undetected amount of energy, whose existence can be inferred from the condition of energy conservation and which indicates the presence of an elusive particle.

Molecule   A building block of matter consisting of several atoms sharing some of their electrons.

Multiverse   The hypothetical set of multiple universes, each characterized by different physical laws, that could simultaneously exist in the physical reality.

Muon   The elementary particle similar to the electron, but about 200 times more massive.

Muon chamber   The set of instruments for measuring the trajectories of muons.

Naturalness problem   The conceptual difficulty posed by the existence of a large ratio between the weak and the Planck lengths, in spite of the natural tendency of virtual particles to erase any difference between the two lengths.

Neutralino   An electrically neutral hypothetical particle predicted by supersymmetry.

Neutrino   The neutral elementary particle that interacts only through the weak force. There are three kinds of neutrinos, associated with the electron, the muon, and the tau respectively.

Neutron   The electrically neutral particle constituting an element of atomic nuclei. It is made of two down quarks and one up quark, held together by gluons.

Nucleus   The dense region at the centre of an atom, consisting of protons and neutrons held together by the strong force.

Photon   The particle that communicates the electromagnetic force.

Pion   The least massive type of meson.

Planck length   (10^{-35} metres) The distance at which quantum-mechanical effects in the force of gravity become important.

Positron   The antiparticle of the electron.

Proton   The positively charged particle constituting an element of atomic nuclei. It is made of one down quark and two up quarks, held together by gluons.

QCD   (Quantum ChromoDynamics) The theory describing the strong force.

QED   (Quantum ElectroDynamics) The theory describing the electromagnetic force.

Quadrupole magnet   A device producing a magnetic field used to focus the proton beams.

Quantum field theory   The theory in which particles are interpreted as lumps of entities distributed in space, called fields, and in which forces are the result of interactions among fields or, in other words, of particle exchanges.

Quantum mechanics   The theory of the microscopic world, which is characterized by an intrinsic uncertainty and an indeterministic nature, and in which particles and waves have a common interpretation.

Quark   An elementary particle that experiences the strong force. There are six kinds of quarks: down, up, strange, charm, bottom (or beauty), and top (or truth).

Quark–gluon plasma   A form of matter at high density and high temperature, consisting of almost free quarks and gluons.

Quench   The process in which a superconductor is warmed up above its critical temperature and becomes resistive, losing its superconducting properties.

Radio-frequency cavity   The device producing an electric field, oscillating at radio frequencies, that accelerates the proton beams.

RHIC   (Relativistic Heavy Ion Collider) The heavy-ion collider operating at BNL since 2000.

Signal   The sample of events that cannot be accounted for by the Standard Model and indicates the presence of new particles or new phenomena.

SLAC   (Stanford Linear Accelerator Center) The US research laboratory established in 1962 and located in Menlo Park, California.

SLC   (Stanford Linear Collider) The linear accelerator colliding electron and positron beams that operated at SLAC between 1989 and 1998.

Slepton   A hypothetical particle, which is the supersymmetric partner of a lepton.

Special relativity   The theory that describes motion, modifying the predictions of Newtonian mechanics when velocities come close to the speed of light.

Spin   The incessant rotation of a particle around its own axis, described by quantum mechanics but incompatible with classical physics.

Spontaneous symmetry breaking   The condition of the state of a system violating the symmetry of the physical laws.

SPS   (Super Proton Syncrotron) The accelerator operating at CERN since 1976 that has handled beams of protons, antiprotons, electrons, positrons, and several kinds of nuclei.

Squark   A hypothetical particle, which is the supersymmetric partner of a quark.

SSC   (Superconducting Super Collider) The proton collider that would have been the largest in the world. Approved in 1987, the project started to be built in Waxahachie, Texas, but was cancelled in 1993.

Stable particle   A particle that does not decay (like the electron).

Standard Model   The theory that describes all known particles and their interactions through electromagnetic, weak, and strong forces.

String theory   A hypothetical theory that consistently describes gravity and quantum mechanics in terms of fundamental entities called strings.

Strong force   One of the four fundamental forces. It is responsible, for instance, for holding together quarks inside hadrons, as well as protons and neutrons inside the atomic nucleus.

Superconductivity   The property of some materials of conducting electric currents with no resistance and of expelling the magnetic field from their interior.

Superfluidity   The property of some materials to flow with no viscosity.

Supergravity   The extension of general relativity that includes supersymmetry.

Superparticle   A particle moving in superspace, which corresponds to two ordinary particles with different spin.

Superspace   A hypothetical space with new dimensions whose coordinates are governed by unusual algebraic rules.

Superstring theory   The version of string theory that includes supersymmetry.

Supersymmetry   The symmetry of a system in superspace, which involves the exchange of particles with different spin.

Symmetry   The property of a system to remain unchanged under a well-defined manipulation.

Symmetry (continuous)   A symmetry corresponding to a smooth manipulation of the system.

Symmetry (discrete)   A symmetry corresponding to an abrupt manipulation of the system.

Symmetry (global)   A symmetry corresponding to a manipulation of the system acting identically at every point in space and every instant in time.

Symmetry (local or gauge)   A symmetry corresponding to a manipulation of the system acting differently at various points in space and instants in time.

Synchrotron radiation   The electromagnetic radiation generated by electrically charged particles whose trajectories are bent in a magnetic field.

Tau   The elementary particle similar to the electron, but about 3500 times more massive.

Technicolour   The hypothetical theory of electroweak symmetry breaking that predicts the existence of a new force in nature, but no elementary Higgs boson.

Technihadrons   New particles predicted by the theory of technicolour.

Tesla   A unit of magnetic field strength (or magnetic flux density).

TeV   A unit of energy equal to one thousand billion electronvolts.

Tevatron   The accelerator colliding proton and antiproton beams operating at Fermilab since 1987.

TOTEM   (TOTal Elastic and diffractive cross-section Measurement) The LHC detector devoted to the study of particles produced along the direction of the LHC beams.

Tracker   The instrument recording the paths of electrically charged particles passing through the detector.

Trigger   An electronic system for identifying potentially interesting collision events that are retained for offline analysis.

Uncertainty (or Heisenberg) principle   The principle of quantum mechanics asserting that there is a fundamental limitation with which two complementary physical quantities (like energy and time, or momentum and position) can be simultaneously known.

Vacuum   The state of a system with the lowest possible energy.

Virtual particle   A particle that, according to quantum mechanics, can exist for a very short time carrying an amount of energy unrelated to its velocity.

W   One of the particles that communicate the weak force.

Wavelength   The distance between two consecutive crests (or troughs) of a wave.

Weak force   One of the four fundamental forces. It is responsible, for instance, for beta radioactivity and for the nuclear fusion processes that make the sun shine.

Weak length   (10^{-18} metres) The distance range of the weak force.

WIMP   (Weakly Interacting Massive Particle) The hypothetical particle constituting dark matter.

WMAP   (Wilkinson Microwave Anisotropy Probe) The NASA spacecraft launched in 2001 that has performed precise measurements of the cosmic microwave background.

Z   One of the particles that communicate the weak force.

Zeptometre   The unit of length equal to a billionth of a billionth of a millimetre (10^{-21} metres).

Zeptospace   The physical space as viewed at lengths of less than 100 zeptometres, which is the goal of the LHC explorations.

Index